P9-DHS-812

DATE DUE

FE 2 [illegible]			
[illegible]			
MR 1 [illegible]	MAY 18 '93		
AP 10 '81	RT'D APR 26 '93		
MY 5 '81			
MR 14 83			
MR 25 '85			
FEB 26 1985			
MAR 22 1988			
MAR 18 1990			
APR 8 '90			

GAYLORD PRINTED IN U.S.A.

Sounds of Music

Sounds

BRITISH BROADCASTING CORPORATION

# of Music

CHARLES TAYLOR

ML
3805
T2.29

Published by the British Broadcasting Corporation
35 Marylebone High Street, London W1M 4AA

First published 1976

© C. A. Taylor 1976

ISBN 0 563 12228 5

Printed in England by Butler & Tanner Ltd
Frome and London

13.00 J-Smith 00043 Physics 12/77

Contents

417422

ALUMNI MEMORIAL LIBRARY
Creighton University
Omaha, Nebraska 68178

Charles Taylor and some of the audience at the Royal Institution

INTRODUCTION

One of the most astonishing of human endowments is our ability to communicate ideas to each other by making tiny changes in the pressure of the surrounding air. The simplest word of a very young child, the eloquence of a great poet, the excitement of a favourite pop tune, or the full glories of symphony and opera are all conveyed to us by sound waves – which are only minute variations in the pressure of the air between the source and ourselves.

When one contemplates the complexity of the information transmitted and the speed with which it is deciphered it is clear that the minutest detail of these pressure variations must be significant. To study them we shall obviously need some device for displaying the variations in graphical form as the sound is being made, and hence quite early in chapter one, after a brief discussion of the ways in which artists and musicians describe musical sounds, we shall meet the cathode-ray oscilloscope. Immediately we have exposed one of the underlying purposes of this book – to show the intricate way in which Science and Art in the study of the Sounds of Music are interdependent; the chart (appendix I) at the end of the book epitomises this relationship. Interdependence between many sciences – physics, psychology, physiology, etc. – will also be a constantly-recurring theme.

In any piece of writing that is intended to convey a considerable body of information to non-specialist readers there is a problem of sequence. At whatever point one begins, one frequently finds the immediate need to refer to something not yet discussed. To avoid this difficulty, and to provide some kind of logical structure – however tenuous – the book has been written round the central theme of the journey which sound waves make from the moment at which they are created until finally they evoke responses in our brains. Chapter one, 'Making and Measuring the Waves', deals with the start of the journey. Like any piece of scientific work, it must begin with some definitions and establish precise ways of describing the phenomena being studied, in this case musical sounds. There are, of course, many ways of describing musical sounds: we can use words, we can draw pictures, we can use the complex notation that musicians have developed over the years, or we can seek to find scientific terms and quantities based on measurement and observation. Whichever method we choose, the problems mount very rapidly as the sound increases in complexity. We shall therefore begin with very simple sounds for which it is relatively easy – both aesthetically and scientifically – to distinguish between meaningful and useless noises. It soon emerges that regularity of the sequence of pressure-changes – the sound waves – in the air is the predominant characteristic of simple musical sounds and this can be created by using objects which vibrate, objects which rotate, by electronic circuits which 'oscillate' or even by computers. Though not aesthetically very attractive, these simple sounds help us to take some basic, but important, steps in understanding the primary scientific requirements of a musical instrument and in finding out the relationships, for various sound waves, between their different effects on our ears and brains and the associated quantities that can be measured scientifically.

It is soon obvious that these sounds – though musical in the sense that they may have a definite place in a musical scale – are very dull compared with the notes produced by real instruments, whether of the traditional orchestral type, or members of the modern family of computers and synthesisers. We shall also find out that speech, even when produced by the most raucous voice, has complex, but readily identifiable musical characteristics; our vocal apparatus is a type of musical instrument even when used for whispering! What is the origin of the quality differences between these real, meaningful tones and the simple sounds with which we started? In chapters two and three we shall begin to look for some answers to this question.

A sound wave does not spring into being 'ready made'; it has to develop, and it turns out that the first fraction of a second in its life is of very great importance in helping us ultimately to identify the sound. In chapter two, 'From Small Beginnings', we shall look at the various ways of starting up sound sources – blowing, bowing, striking, plucking, etc. – to see how the particular method adopted affects the quality. The consonants in speech must be included under this heading and we can learn a great deal from comparisons between sounds produced by the voice and sounds produced by man-made devices. Most of the primary sources of sound are too weak on their own and need to be reinforced or amplified. The body of a violin, for example, amplifies the sound produced by the string which, on its own, would be almost inaudible. But, even when the bow has made the string vibrate, the string must communicate its energy to the body and this takes time; the body is reluctant to vibrate and, during the time taken to persuade it to vibrate in sympathy with the string, the sound produced is significantly different from the steady note which finally emerges. The way in which the note begins is greatly influenced by the skill of the player as well as by the characteristics of the instrument itself.

In chapter three, 'Growing and Changing', we shall follow still further the growth of the note. The need for amplification leads us to discover one of the vital factors which determine why one particular instrument sounds different from another of the same type; for example, why a Stradivarius violin sounds totally different from a cheap copy, even when played by the same player. The amplifier – in this case the violin body – does not behave perfectly; that is, it does not merely make the sound louder but it changes it in the process, and it does not even change each note in the same way. There is thus a characteristic 'signature' or 'finger-print' which may identify the music produced by a given instrument. The effect applies to all instruments and most especially to the human voice. The basic notes produced by the vocal chords are amplified and modified by all the various throat and nose cavities to distinguish between vowel sounds and also between individual voices. Sounds produced by different instruments are never constant; they grow and die away in different ways and the pattern of growth and decay offers further information which helps us to interpret the sounds.

Up to this point the discussions may have seemed somewhat abstract and, though various specific instruments have been mentioned, the detail of their construction and the importance of the instrument maker's contribution to the final sound may not have emerged. Chapter four, 'Craftsmanship and Technology', considers first the remarkable way in which the design and construction of instruments has developed over hundreds of years with little or no help from science. Recently, however, science has begun to make contributions and some of the most sophisticated tools of the physicist – holography and digital electronic systems among others – are being used in attempts to improve traditional instruments, to produce totally new sounds and to understand the mechanism of the human voice. The rapid advances in technology – especially electronics – have also led to the production of a whole new family of musical instruments – electronic synthesisers – which enable us to create sounds in a much more precise and controllable way than with traditional instruments. Whether this is a desirable development and whether the resultant music is acceptable is a subject of burning controversy but, undoubtedly, electronic instruments are here to stay. Whatever our views of their aesthetic value at the moment, we can learn so much from them, both about the scientific and the musical aspects of sounds, that it is worth spending some considerable time on them.

The sound waves are now established in the

air, but there are many hazards ahead before they reach their final destination. Chapter five deals with the various modifications that occur 'On the Way to the Ear' which may well make the waves which actually impinge on the ear drum surprisingly different from those which left the instrument. All sounds are produced in some kind of environment – a room, a concert hall or in the open air – and the patterns of waves set up are different in each case. The problem of designing a suitable environment for concerts, for speech or for other purposes has long been of concern to architects; the physics of the behaviour of sound waves in various environments is well understood, but because an environment is never designed for one very specific use, compromises are always necessary in applying the principles. This means that judgements must be made. Physicists are well able to help with the measurements and predictions of likely behaviour, but the criteria by which they are assessed depend on sociological, psychological, aesthetic and, of course, economic factors. Nowadays musical sounds travel towards us via many routes in addition to those through the air; transistor-radios, hi-fi record reproducers, tape-recorders and public-address systems are all familiar and all imprint their own characteristics on the waves.

Finally, in chapter six, we consider 'The End of the Journey', when the waves arrive at our ears. The hazards are by no means over, however, and in many ways the last stage can be considered the most complicated, the most fascinating and is probably the least well understood. The purely physical mechanism of the middle and inner ear introduces some changes, but the way in which the brain deciphers the information is clearly extraordinarily complicated. If we listen to a full chorus and orchestra in some operatic climax the whole of the information is conveyed as a single wave; the pressure in the air just in front of the ear has only one value at one time. Yet somehow we are able to disentangle the information and listen at will to the soprano, the first violin, the coughs of the audience, or the rustling of a toffee paper in the row behind. We recognise them all separately by using all kinds of small clues which are collected, catalogued and correlated in tiny fractions of a second. Why do some combinations of tones seem pleasant and others discordant? Why do some tunes make us want to laugh and others bring us to the verge of tears? What causes the tingle in the spine that many of us feel when some particularly moving performance of a great masterpiece is experienced? Can we always believe our ears or are there aural as well as optical illusions?

These are some of the questions we shall raise, but only partial answers will be given to most of them as this is an area where there is still much controversy and discussion; the mechanisms are not clearly understood and physics alone will not decipher them.

One may well ask whether the title of the chapter is really a fair one. Would it perhaps be truer to say that the arrival of the sound waves at the ear drum really represents the beginning of another and even more complicated and exciting journey! The fact remains, however, that our ears and brains do perform the miraculous transformation from tiny pressure variations in the air to the splendour of music, and our study of the journey of the waves may have contributed something to our understanding both of the part played by physics in the study of the Sounds of Music and of its limitations.

C. A. TAYLOR
Department of Physics, University College, Cardiff

1.1 a–d *Cartoonists' descriptions of musical sounds; (a), (b) and (c) are by Hoffnung and (d) is by Steinberg*

MAKING AND MEASURING THE WAVES

DESCRIBING SOUNDS

Every day of our lives we use our voices to create sounds with which to exchange ideas with other people round about us, and, though you may not have thought about it in this way before, those sounds are highly musical. Even the most harsh, discordant voice can be analysed into components which have recognisable musical characteristics, as we shall see later on in the book.

Words themselves are made up of sequences of musical sounds and, in a paradoxical kind of way, we can use words in turn to describe musical sounds. That is where, in fact, we must begin. How does one satisfactorily describe a musical sound? Suppose we take a small bottle and blow across the top – how would one describe the sound? Hoot? Toot? Perhaps these words would do – but they are ambiguous and might equally well represent a motor horn or an owl calling to its mate. Words like 'Doi-oing' conjure up the kind of sounds used in cartoon films and 'ping' or 'ting' are fairly specific. Clearly, though, there is a total lack of precision in verbal description.

Stan Barstow, in his novel *The Watchers on the Shore*, produces a remarkable verbal description of a piece of music. It occurs when Albert has just returned to his flat and joined Conroy and they decide to listen to a record of the *Roman Carnival* overture by Berlioz.

'. . . and climbs to a climax that's all snapping, snarling brass . . . *doo-ah rratatah dee doo doh dah, rrumdidumdidumdidumdidumdumdah dooooh daaaAH!*'

If you know the overture, this is a brilliant and precise evocation; if you do not, then it could mean all kinds of different things. As far as the subject-matter of the present book is concerned, that initial phrase 'If you know the overture . . .' is almost prophetic; the idea that memory or experience is needed in almost any kind of perception or interpretation of even the simplest sound will constantly recur.

Various cartoonists have attempted very successfully to express sounds in a visual way and 1.1 shows four of my favourite examples. Though one has no difficulty in imagining the sound – this is almost hearing with the eyes – there is still little precision. One could hardly imagine an orchestra playing from a score which consisted of a sequence of cartoons with or without words! They would, no doubt, have great fun, but the result would be somewhat unpredictable.

Musicians on the other hand have developed very sophisticated and precise ways of describing musical sounds. 1.2, for example, is a fragment of a thirteenth-century motet. There are three voices singing not only different melodic parts

1.2 *Thirteenth-century motet 'O Maria Davidica'*

1.3 *Last twelve bars of 'Symphonie Fantastique' by Berlioz*

but also different words; but 700 years later we can still interpret the sounds and be reasonably sure that what we produce resembles very closely what the composer intended.

Already at that time the basic idea of black dots placed on sets of lines to represent the movement of the melodies is established. 1.3 shows a complicated example of a score still using the basic idea of dots and lines to represent both timing and pitch. It is a fragment of Berlioz's Fantastic Symphony and the extract shown lasts about eight seconds. Most of the instruments of the orchestra are involved and right at the very end the composer had to squeeze in an extra line at the bottom of the page for a cymbal clash. Although this looks quite complicated we shall see later that every single note on the page itself corresponds to a pretty complicated and very rapid series of changes in the pressure of the air and to describe *one* of these notes on *one* instrument lasting say one second in precise enough terms for it to be synthesised on a computer – though this is really jumping ahead to chapter four – might involve anything up to 40,000 separate numbers! Truly the musician's score is an effective and condensed way of describing sounds. But of course it depends on the knowledge and experience of the performer; if in thousands of years' time all records and descriptions of existing instruments had been lost a future researcher might conceivably deduce roughly what the melody would sound like, but the word 'violin' or 'oboe' would not help him to reconstruct the quality of tone at all.

During the twentieth century musicians themselves have extended the technique of scoring still further and in 1.4 and 1.5 we see an extract from the manuscript score of 'Time Sketches III', by my colleague Keith Winter. The piece involves a combination of conventional instruments used in conventional ways together with electronic sounds pre-recorded on tape and live bass modified electronically. In 1.4 we see the parts for bass and electronics and in 1.5 the more conventional component of the same few bars. This illustrates well the extension of scoring technique and one

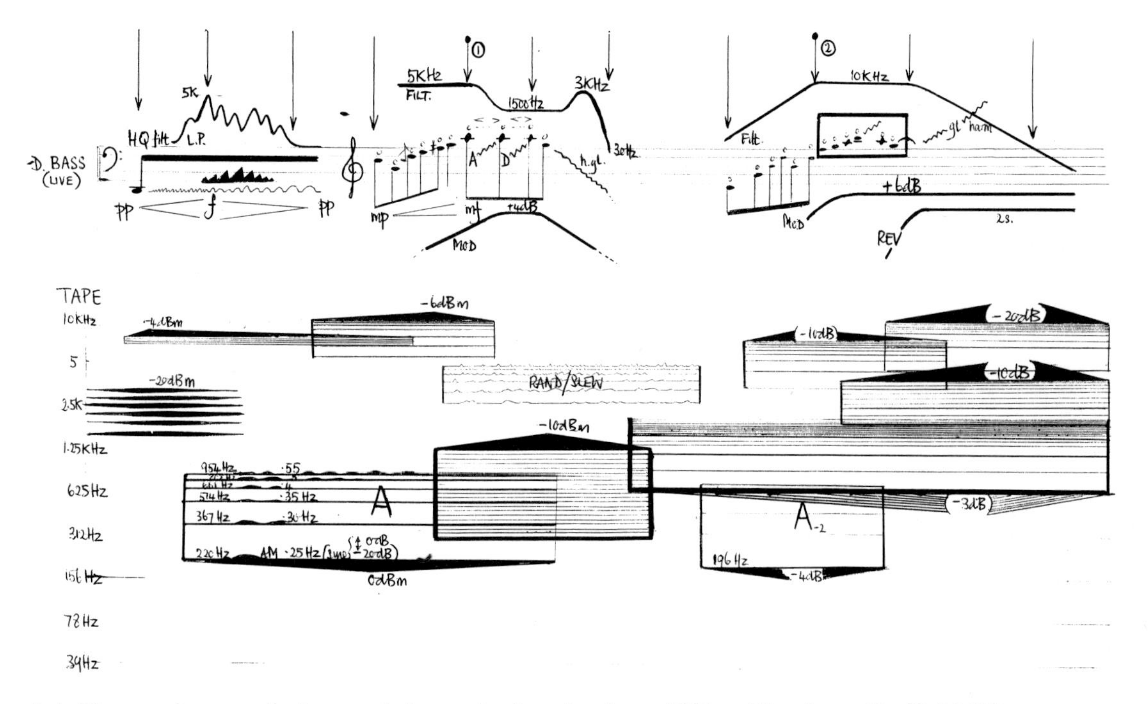

1.4 *Manuscript parts for bass and electronics for a few bars of 'Time Sketches III' by Keith Winter*

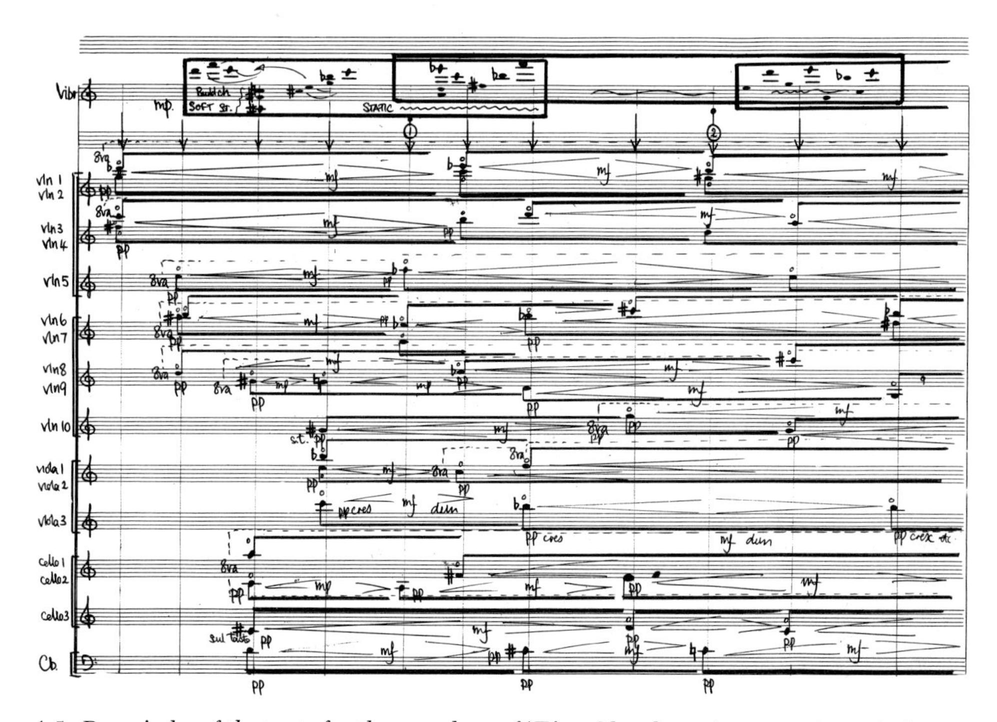

1.5 *Remainder of the parts for the same bars of 'Time Sketches III' as were shown in 1.4*

can see how the composer has introduced scientific quantities, such as Hertz (Hz), decibels (db), etc. (which for the moment may be said to indicate pitch and loudness), in order to gain sufficient precision in describing new kinds of sounds.

WHAT IS A SOUND WAVE?

To be completely scientific we need to go much further even than Stockhausen; but before we can plunge into that problem we must stop to ask ourselves some questions about the nature of sound itself. We have already indicated that sound involves pressure variations in the air; what happens when, for example, we speak is that we create a complicated series of compressions and expansions in the air and these then travel outwards as a succession of waves towards the hearer. It is important to be quite clear what we mean by a wave. It is not *necessarily* something which changes up and down in a regular sort of way, as the layman might imagine from the everyday use of the expression 'wavy'. Scientifically speaking the essence of a wave is that *something* travels along from one point to another without a wholesale transfer. For example, most of us have sometime or other set up a row of dominoes and, on pushing one over, all the rest fall down in succession. The key points are first that the double-six in 1.6 has not moved to the position of the double-one. But *something* has moved along the row, and if we had a little switch which could be knocked down by the double-one we could operate a light from a distance using energy passed along the row as a wave. Secondly it takes time for the wave to travel from the double-six to the double-one – and we shall refer frequently to the importance of this kind of time delay.

Our prime concern is with sound waves, which also take time to move from one point to another. If I click my fingers I create a disturbance of the air which travels as a wave at about 700 mph or 300 metres per second. In a normal-sized room this means that the time delay is so small that it is not noticed (as a matter of interest work out for yourself how long it would take for a sound wave to move from where you are sitting to a wall in front of you and return to your ear); we are all familiar, though, with the delay between a flash of lightning and the arrival of the sound of the thunder from a mile or two away, or between seeing a woodman's axe strike a tree on the other side of a valley and hearing the sound. A more realistic model of a sound wave than the dominoes is the toy, the 'slinky spring' shown in 1.7. The compression wave created by pushing the end coil inwards and releasing it travels along and again takes time, though the spring as a whole has not moved.

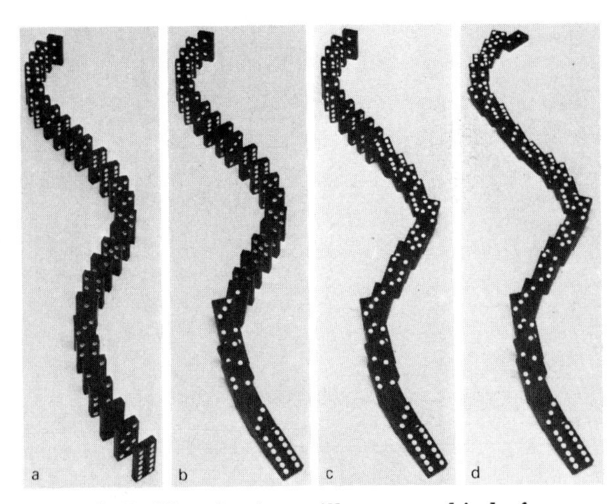

1.6 a–d *Falling dominoes illustrate a kind of wave*

1.7 a–f *Wave travelling along a spring taken at intervals of approx. 1/10, 1/5, 1/3, 2/3 and 1 second*

CAN WE SEE SOUND WAVES?

It would be very useful if we could examine the detailed motion of this spring rather more closely, and this can be done by adding a simple lever and a mirror at the end. 1.8 shows diagrammatically how this is done, so that a beam of light falling

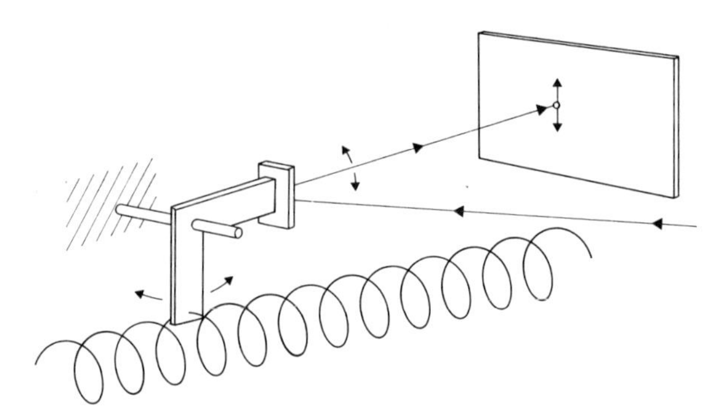

1.8 *A method of converting the movement of waves along the spring into vertical motion of a spot of light*

on the mirror is reflected on to a screen and dances up and down along a vertical line in step with the arrival of the waves. We can examine the detailed variations of the pressure of the air in a sound wave by very much the same sort of technique. We use a microphone which changes the variations in pressure of the air as the wave arrives into variations in an electric current which, in turn, can be made to deflect up and down the spot of light on the screen of a cathode-ray oscilloscope. This is perhaps the most basic and useful tool that physics has given to students of sound waves and we shall see many examples of its use as the book progresses. 1.9 a and b show the undisturbed spot and the line produced by the vertical movement of the spot when a wave is received. This is still not as useful as it might be because the movement is all in one vertical line. 1.10 shows how we

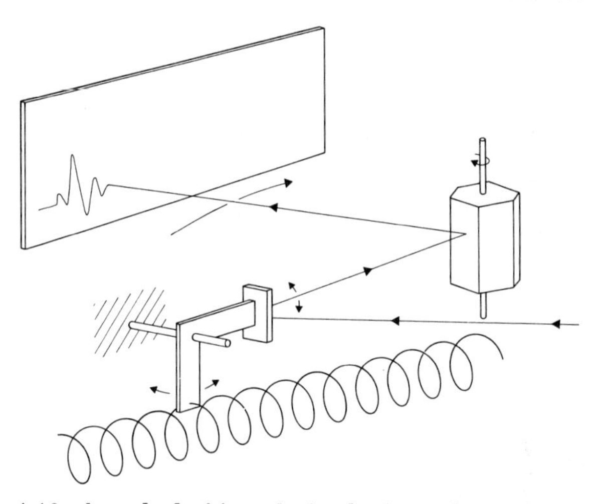

1.10 *A method of introducing horizontal scanning in the light-spot experiment of 1.8*

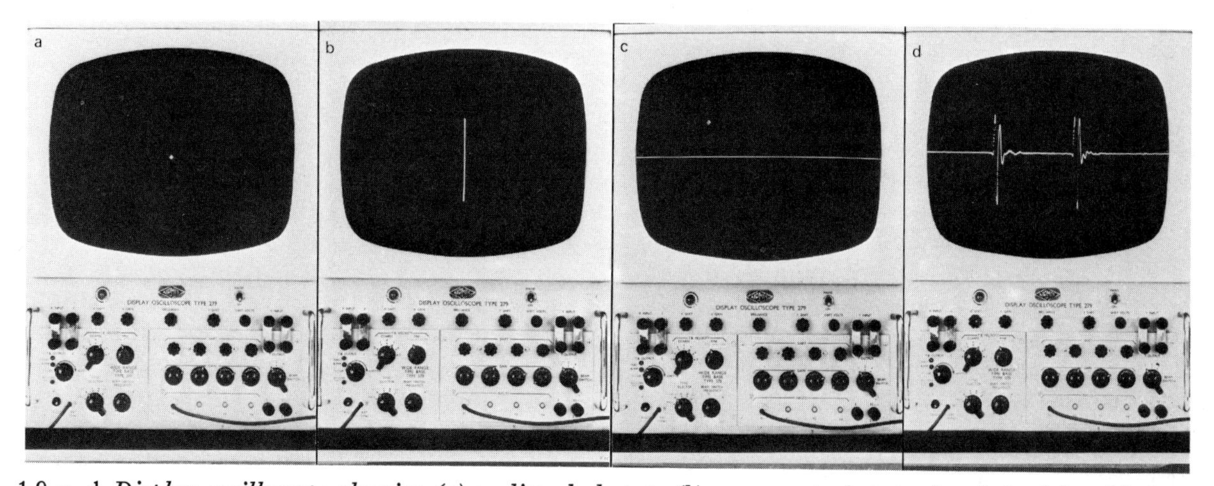

1.9 a–d *Display oscilloscope showing (a) undisturbed spot; (b) movement of spot when fed solely with input from microphone; (c) movement of spot when fed solely with scanning voltage; (d) combined movement when fed with scanning voltage and with microphone receiving clicks*

can easily change the motion of the light spot from the spring into a graph by letting the spot fall on a rotating drum covered with strips of mirror. Each strip picks up the spot in turn and spreads it along a line and, if the mirror is moving fast enough, the eye sees the result as a graph. In the early days of sound research such a rotating mirror system for 'scanning' was actually used to study the movement of a diaphragm in response to sound. Nowadays, thanks to electronics, we can simply switch on the 'scan' on the cathode-ray oscilloscope to produce a much more effective representation of the wave. 1.9 c and d show the screen of the oscilloscope first with the scan switched on and then responding to the arrival of waves produced by clicking the fingers in front of a microphone. Sometimes the spot of light is allowed to fall directly on to photographic paper and then the trace is black instead of white (see, for example, 1.12); if one is really pedantic this apparatus is an oscillo*graph.*

Now we have a precise scientific way of looking at the graph of the pressure variations against time and it would be interesting to compare the wave patterns resulting from various sounds to see whether we can distinguish the characteristic that identifies them as musical or that distinguishes one kind of music from another. 1.11 shows traces of four different sounds. We have deliberately printed the caption, telling you what each is, on a later page (page 13) so that you can try for yourself to identify the traces without any help. It is very difficult indeed and, even after years of experience, it remains very difficult. Of course we are not really playing fair because the total time represented by each trace is less than half a second and in that time even the ear might find some difficulty. But even when much longer sections of the trace are presented to the eye it is very difficult to recognise complex sounds like this. At the end of the book we have included an oscillograph trace lasting about twenty-five seconds and this – though representing a relatively simple piece of music – shows how, in any passage in which more than one instrument is playing, the wave form becomes highly complicated and changes rapidly with

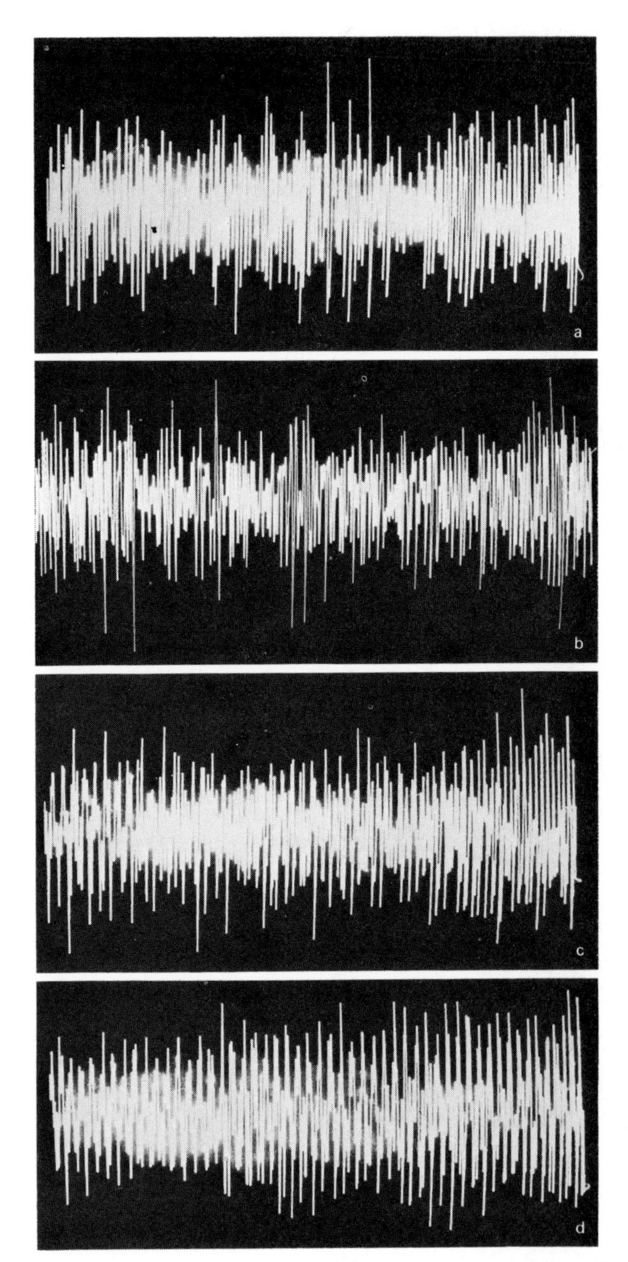

1.11 a–d *Oscilloscope traces of various sounds all recorded at the same speed; the time for each trace is about half a second. The identity of the sounds is revealed on page 13*

time. By now you will undoubtedly have looked up the answers and found that one of the traces is that of an orchestra tuning up. Is this music or is it noise? This question really highlights the sort of difficulty one is continually meeting in this field of study; clearly the sounds are musical, since they are made by musical instruments, but the net result is not very pleasant except that it may have pleasant psychological effects as it whets one's appetite for the concert which usually follows!

THE CHARACTERISTICS OF SIMPLE SOUNDS

This experiment, then, does not take us much nearer to answering the question of what, scientifically speaking, characterises a musical sound. We are dealing with sounds that are far too complicated and, as so often happens in a piece of scientific research, we must begin by making great simplifications; later we shall be able to move slowly back to the complexities of the full orchestra. Mendelssohn is very helpful in providing us with a transition; at the end of the first movement of the Violin Concerto a loud and complex chord dies away, but the bassoon alone of all the instruments lingers on to provide a bridge to the second movement (a few bars of the score are shown in 1.12a). 1.12b–f show the oscillograph trace of this transition and one immediately sees the steady regular repetition which is characteristic of a long note on a single instrument. This is still not the kind of graph that one usually associates with a simple wave and the actual shape that repeats is quite complicated. If, however, we move to a really simple instrument we can achieve a reasonable approximation to what the scientist would call a pure tone. 1.13 a is the trace of a treble recorder playing 'C'. Now, at last, we have a wave which looks something like the layman's idea of a wave. Using the treble recorder it is easy to demonstrate some of the simple relationships between the graph of the pressure variations and the effect on the ear. If the recorder is played louder the up and down movement or amplitude of the waves increases – in other words the amount of pressure variation increases although the rate at which it happens does not change. If the note being played is moved up an octave the number of wave crests in a given time is doubled, or, scientifically speaking, the frequency of the waves

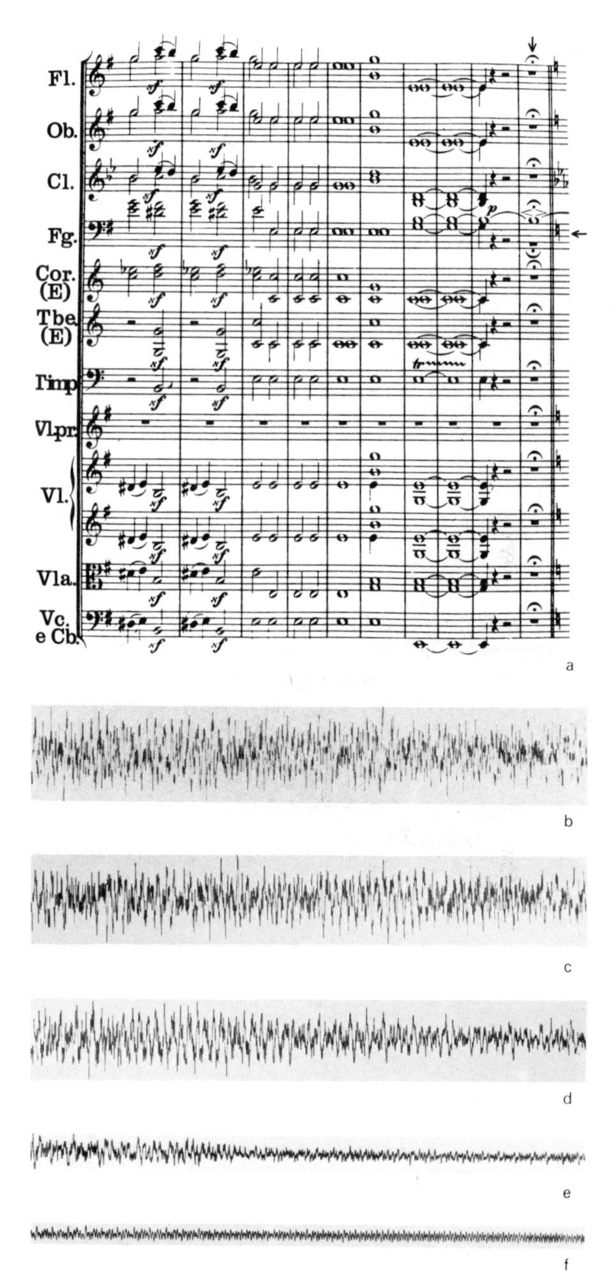

1.12 *(a) Last nine bars of the first movement of Mendelssohn's Violin Concerto and the 'transition' bar to the second movement. (b)–(f) Oscillograph trace of the final chord and the transition to the bassoon. (A different kind of recording device was used here: it produces a moving dot of ultra-violet light on moving photographic paper and hence the trace is black on white*

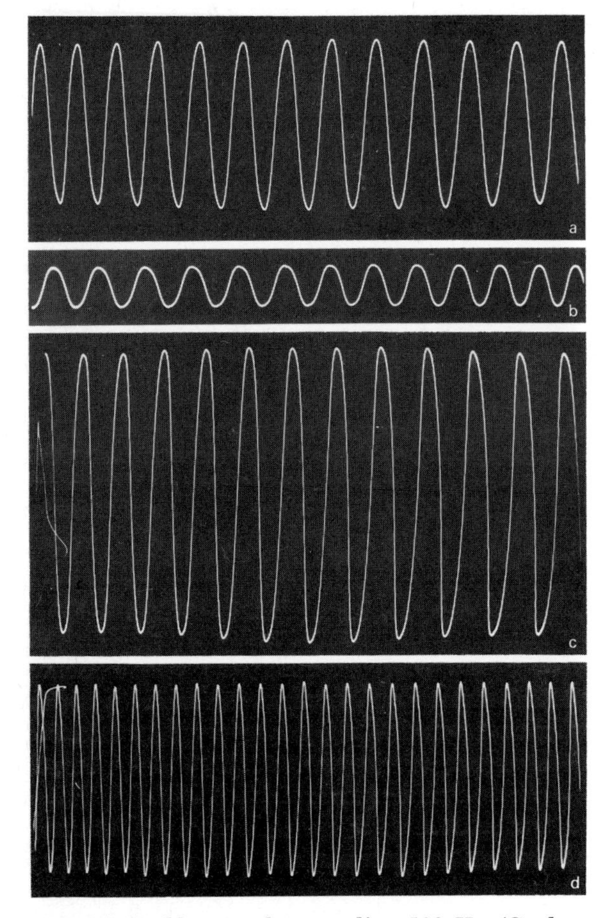

1.13 *(a) Treble recorder sounding 523 Hz (C above middle C). (b) As (a) but softer. (c) As (a) but louder. (d) Treble recorder sounding 1046 Hz (octave above (a)). All these traces last about 25 milliseconds (ms)*

is doubled. We shall see in chapter six that this is an over-simplication, but we need not worry too much about that now. Both relationships are illustrated in 1.13 b, c and d.

SOME FACTS AND FIGURES

To round off this simple introduction two obvious questions might be asked and answered. The first concerns the actual magnitude of the pressure variation, and it turns out that between the high and low pressure points in the wave about 5 yards away from a treble recorder playing C fairly loudly the pressure difference is something like 0.000015 lb per square inch which, compared with atmospheric pressure, amounts to 1 part in a million! The second concerns the number of crests per second in the wave and, again for the note C above middle C, this turns out to be 523 per second. Nowadays physicists would describe this as a frequency of 523 Hertz (Hz). It sometimes comes as another surprise to non-scientific instrumentalists to realise that they are producing so many compressions per second!

To the physicist – who is always thinking of new questions – these two answers lead us to ask what the limits of detection by our ears are, both in pressure difference and in frequency. A fuller answer to these questions is given in chapter six and in appendix III, but straight away we can reveal the astonishing fact that for average sounds and the average 'man-in-the-street' there is a ratio of something like 1,000,000 to 1 between the pressure difference in the wave for a sound that makes our ears begin to feel pain and a sound we can only just hear. The quietest sound (the proverbial pin dropping) would correspond to a pressure change of 2 or 3 parts in 10,000,000,000 and the loudest we can stand without pain would be 2 or 3 parts in 10,000.

This is, of course, an enormous range and we can see immediately what astonishing measuring instruments our ears are. Just compare with the kind of measuring instrument found about the house. With a foot ruler we can measure fairly accurately down to about 1/20 inch and up to 1 foot – a range of 240:1; with a kitchen scale we can weigh down to say 1/2 oz and up to say 5 lb – a range of 160:1; with the tyre pressure gauge used on a car we might measure down to 1lb per square inch and up to 100 lb per square inch – a range of 100:1. Of course, we could measure wider ranges if we used several different instruments, but the ear – one single instrument – has a range of 1,000,000:1!

But what about frequency? Again this is covered in more detail in chapter six and appendix I, but, broadly speaking, most people can hear as a musical tone anything vibrating at more than about 20 times per second or 20 Hz and

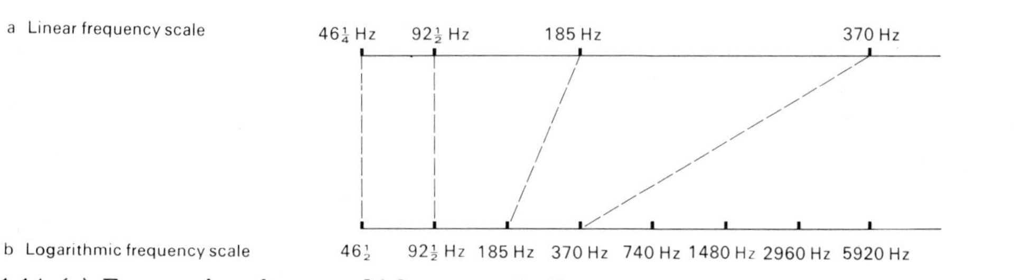

1.14 *(a) Frequencies of notes which are musically one octave apart represented on a linear scale (length proportional to frequency). (b) Frequencies of notes which are musically one octave apart represented on a logarithmic scale (each octave the same length)*

less than about 20,000 Hz. The exact limits vary a good deal from person to person and also change – especially at the top end – with age.

FREQUENCY AND OCTAVES

Because of the large range in both pressure variation and in frequency to which our ears are sensitive we often find it convenient to use a special form of scale, particularly when drawing graphs and charts. It is most easily introduced by considering the frequencies of notes an octave apart. If we start off with a low note and let it glide upwards there are numerous points at which we experience the sensation that it would be natural to stop. This is especially true if we go on playing the lower note as well. If you are the kind of person who habitually describes himself as 'tone deaf' try this simple experiment. Choose any note near the bottom of the piano keyboard and play it together with the next one above it – continue to play the lower note and move the higher note up one at a time (using both black and white notes). You will certainly find some of the two-note chords pleasanter than others and, in particular, every time you play notes which are 12, 24, 36, etc., apart you should experience a strong sensation of coming back to rest. Frequency measurement on these notes – the notes are one octave apart – shows that they represent a doubling of the frequency. For example, suppose you happen to start on the note which your musical friends tell you is the lowest F♯ on the piano; your sequence of chords will be F♯G, F♯G♯, F♯A, F♯A♯, F♯B, F♯C, F♯C♯, F♯D, F♯D♯, F♯E, F♯F, F♯F♯. If we measured the frequency of the lowest F♯ it might be 46¼ Hz and the next one up would then be 92½ Hz – twice the lower frequency. If this experiment continued the next F would be 185 Hz and the next (the one just above middle C) would be 370 Hz and so on.

In other words every time we go up, musically speaking, one octave, the frequency is doubled. 1.14 a shows how a straight frequency scale would appear – the octaves are further and further apart. We therefore decide to use a scale which makes the *musically* equal steps – the octaves – equal *distances* apart and the result is 1.14 b. This is a special type of scale which physicists and mathematicians call logarithmic and it crops up whenever we talk about the relationships between things that can be physically measured (in this case frequency) and things that are matters of perception (in this case pitch). (The scale is in fact the same type as the one that appears on a slide rule.)

The chart of appendix I, which shows the relationship between pitches, frequencies, musical notation and the ranges of various instruments, is arranged on this kind of scale.

■ Sounds corresponding to oscilloscope traces on page 10:

a *Climax of First Movement of Mendelssohn's Violin Concerto*

b *Orchestra tuning up*

c *Chatter of audience entering theatre*

d *Track from Pink Floyd LP 'Meddle'*

MUSIC AND NOISE

Now we are in a position to begin to make some simple sound waves which may or may not be musical. When we first talked about the cathode-ray oscilloscope we saw the shape of the wave produced by a single click or snap of the fingers (1.9 d) – what a physicist would call a single pulse. Suppose now we produce a large number of such pulses not in any regular kind of way; for example we might pour a stream of rice grains from a bottle on to a hard table, listen to applause, or turn up the volume on a radio not tuned to a station.

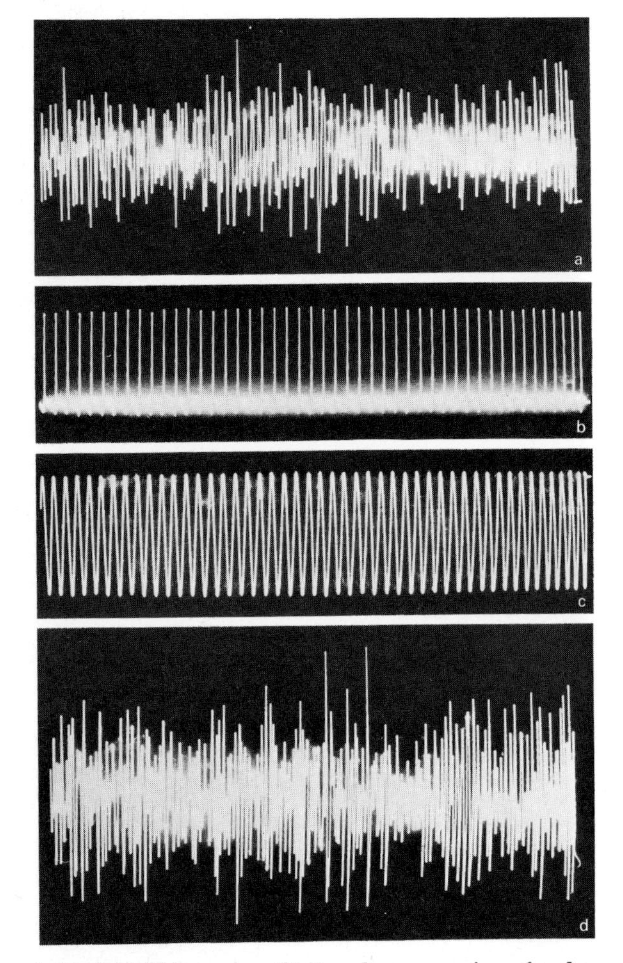

1.15 *(a) White noise. (b) Regular succession of pulses 100 Hz (500 ms long). (c) Pure tone of 100 Hz. (d) Repeat of the portion of the Mendelssohn Violin Concerto from 1.11 a*

It is the sound a physicist would call 'white noise' and its pressure graph is like 1.15 a – just a random mixture of pulses. But suppose we now make the pulses follow one another at regular intervals? This could be done by holding a piece of card near the teeth of a rotating toothed wheel; every time a tooth hits the card we hear a click or produce a single pulse and the resulting pressure graph looks like 1.15 b; if the pulses follow each other fast enough we hear a harsh but quite easily recognisable musical note. It turns out that the pitch would be the same as that of the simple tone whose pressure graph is shown in 1.15 c – but this last one sounds very smooth and not a bit harsh. Thus we see that in its simplest possible form music involves regularity, and noise irregularity. But notice a rather odd thing; just compare the pressure graphs 1.15 a and 1.15 c with those of 1.11. Do you notice that even the most musical (in my opinion, the Mendelssohn) of those – 1.11 c (we have repeated it as 1.15 d) – looks more like 1.15 a (white noise) than 1.15 c (a pure tone)? This merely illustrates once again how difficult it is to get very far merely by looking at pressure graphs when we are dealing with very complicated waves and, for the moment, we must just say that 1.15 d and 1.15 a resemble each other in complexity but certainly not in detailed form – though we do not know enough physics yet to understand what that really means. By the time we reach the end of chapter six we might have found a few clues. For the time being, however, we will content ourselves with using the simple ideas we have developed so far to construct some simple instruments for making musical sound waves and for playing tunes.

SOUNDS FROM ROTATING OBJECTS

The whine of machinery, the harsh note of a circular saw, the wail of a siren are all common examples of sounds of definite pitch derived from rotation. Anything which goes round and round will tend to produce the same succession of clicks, taps or squeaks every revolution and hence, if going fast enough, the regular waves needed to make a musical sound. The siren is

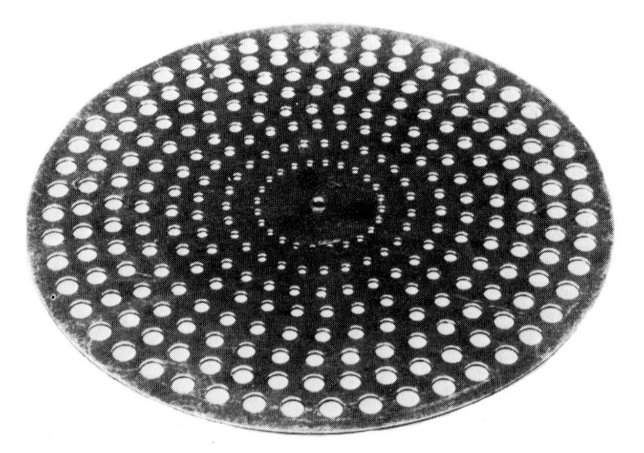

1.16 *Perforated disc to be used as a siren. The eight rows produce the eight notes of one octave of a diatonic scale. The numbers of holes in successive circles are: 24, 27, 30, 32, 36, 40, 45, 48. The ratios between the notes given by any two rows remain constant, though their absolute frequency or pitch depends on the speed of rotation*

one of the very early applications of this principle and in its simplest form consists of a disc with regular circular rings of holes drilled in it; this is spun either by a motor or by hand with suitable gearing and a jet of air is directed on to a particular row by means of a rubber tube and a tapered jet. Mouth pressure is quite sufficient to produce a definite, though quiet, sound, and by blowing at the right moment and moving the jet it is quite easy to play tunes. 1.16 shows a typical disc and gives the numbers of holes in each row needed for a simple diatonic (white notes only) scale of one octave.

A more sophisticated version uses a beam of light (e.g. from a laser, though this is not necessary) which falls on to a photo-cell and whose path is interrupted by the disc. This particular device is mentioned because by varying the shape of the holes one can make the intensity of the light vary instead of merely chopping it off and on, and hence much more complicated waves can be produced. Two well-known types of electronic organ – the Compton Electrone and the Hammond – use rotating devices as the primary tone generators. The Electrone uses changes of electrical capacitance and the Hammond changes in a magnetic field.

SOUNDS FROM ELECTRICAL OSCILLATION

Everyone has experienced at some time or other the dreadful howl that occurs in a public address system if the volume control is turned up too high. Let us think for a moment how this may arise. Suppose that a tiny click, cough, rustle of toffee paper or any other sound occurs accidentally and is picked up by the microphone. The amplifier then feeds the sound to the loudspeaker, thus creating a larger sound wave which travels out to be picked up again by the microphone and the whole cycle keeps on being repeated. The key point is that it does take a certain amount of time for the sound to get from the microphone through the amplifier to the loudspeaker and then from the loudspeaker through the air back to the microphone. This total amount of time will remain the same unless anything is moved or the amplifier is altered, and so what began as a single accidental sound is turned into a succession of sounds at regular intervals that will go on for ever unless the amplifier is turned off. The sound is thrown back and forth from microphone to loudspeaker and we say that the system is oscillating. If the volume is turned down so that the first sound picked up is not reproduced in the loudspeaker loud enough to be picked up again by the microphone, no oscillation occurs. Notice how important the *time* is; it will, in fact, determine the pitch of the howl. If the microphone is taken a long way from the loudspeaker or a large amount of extra cable is put in, the pitch will go down because the time delay is longer and the frequency decreases.

This is a very clumsy way to produce oscillation and it is far simpler to use an amplifier without a loudspeaker or microphone and to feed a little of the output back into the input *electrically*. Such a device – known simply as an oscillator – forms the basis of many electronic music generators – the pitch may be changed by altering the electrical components in the circuit. Insertion of a large capacitance, a large resistance or a large inductance will in general introduce delays within the system and hence produce lower frequencies. We shall discuss the use of this kind of circuit in more detail when the subject of the electronic synthesis of music arises in chapter four. Of course such an oscillator produces primarily an oscillation in an electric current and, in order to turn it into audible sound, the current must be fed to a loudspeaker which converts the electrical changes into movement of a cone or diaphragm, which in turn sets up the necessary waves in the air.

SOUNDS FROM MECHANICAL OSCILLATION OF A SINGLE VIBRATOR

Traditionally musical instruments have always involved mechanical oscillation of some sort and so we shall spend rather more time at this stage studying mechanical systems: some of the ideas developed will, however, play an important part in our later discussions of synthesis and reproduction of music.

The simplest example of a mechanical oscillation is a child bouncing a ball – we are very familiar with the idea that if it is bounced from a height of say 2 feet the regular beat will be slower than if the hand is dropped to 1 foot from the ground. Just as in the last section the sound was thrown back and forth between the microphone and the loudspeaker, so here the ball is thrown back and forth between the hand and the floor, and the time taken is the all-important factor.

Let us now return for a moment to the slinky spring in 1.7. It would be quite possible to imagine a single compression travelling back and forth along the spring from one end to the other and hence setting up the regularity we need for a musical sound if only it could travel fast enough. Take a solid metal rod (brass is good), say about a metre long, and suspend it horizontally by two pieces of string and the experiment can be done. Hit the end of the rod hard with a hammer and it will start to swing, but it will also emit a high-pitched ringing note which is exactly the result of the compression caused by the hammer blow travelling to the other end of the rod and back again. It will travel 2 metres between each return to a given end and so a succession of pulses will be sent out into the air from each end. It turns out that sound travels roughly 3400 metres per second in brass and hence it will travel the 2 metres 1700 times each second and this will be the frequency of the note produced – not far off the A in the third octave above middle C.

An alternative method which produces the note in a more sustained way involves wearing a pair of old cotton gloves which have been liberally covered with powdered resin. The rod is held between the thumb and forefinger of one hand precisely at its mid point, and the rod is stroked with the thumb and forefinger of the other hand, starting about midway between the centre and one end, and moving outwards towards the free end. The rod must be free from grease and it is then usually possible to produce a loud note with very little effort and without gripping tightly with the moving hand. This is an example of an interesting process – to which we shall return in chapter two – known as stick-slip motion. The resin makes the fingers stick to the rod and it is slightly stretched – almost immediately the rod overcomes frictional forces and shrinks again, slipping past the glove in the process. Once again the fingers grip and this process continues. But each 'stretch' travels as a wave to the end of the rod and back and the whole rod is subject to the succession of waves travelling up and down. As we shall see in the next chapter the time taken to travel up and down dominates the process and reacts back on the timing of the stick-slip motion at the glove.

The sound produced by this system, though persisting for a long time, is relatively quiet. This is not really surprising when one remembers that the waves in the air are being created by the to-and-fro movement of each end of the rod and the area in contact with the air is very small. A way of demonstrating that this is indeed so and a very beautiful demonstration of the fact that sound really does travel through the air is provided by an experiment described by John Tyndall as long ago as 1863. Metal discs some 3 or 4 inches in diameter are added at each end of one of the brass rods and it is found that a very loud sound indeed can be produced by the stroking technique. (Two discs prove to be necessary to ensure symmetry so that the waves 'bounce back' in the same way at each end.) A thin sheet of rubber stretched tightly on a circular embroidery frame 6 to 9 inches in diameter has sand scattered lightly on it and the rod is held vertically with its lower disc about 6 inches above the rubber sheet. When stroked to produce a vigorous note the sand dances violently and collects in a series of concentric circles, showing that the vibration of the rod must have been transmitted through the intervening air to the rubber diaphragm. 1.17 is one of the original illustrations from Tyndall's book *Heat: a Mode of Motion.*

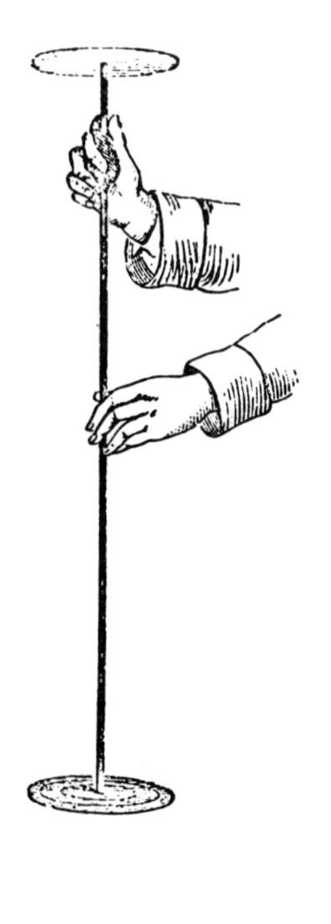

1.17 *Diagram reproduced from John Tyndall's* Heat: a Mode of Motion, *first published in 1863. The brass plates communicate the longitudinal vibrations of the brass rod when stroked to the air and the resulting sound waves set the drum vibrating to produce patterns in the sound. There are also patterns on the brass plates. (Page 231 in 3rd edition, Fig. 69)*

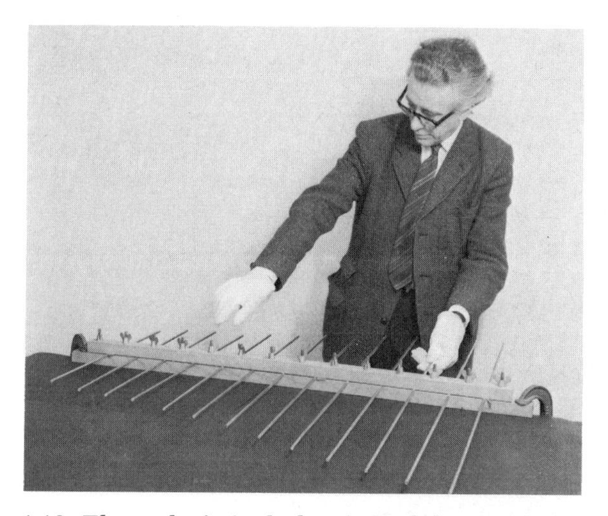

1.18 *The author's 'rodophone'. Each brass rod gives a different note when stroked with resined gloves*

The rod – however it is made to vibrate – provides an example of the class of musical instruments which have one vibrator for each separate note; we could call this family number one. If you wish to play tunes on a member of this family it is necessary to have a number of different vibrators. A series of brass rods of differing length can, for example, be made into an instrument capable of playing tunes if they are all clamped firmly at the mid point on a wooden frame (1.18). The conventional members of this family are the piano (which is really eighty-eight separate instruments each called into play when the appropriate key is struck), harpsichord, harp, organ, xylophone, tubular bells, etc.

It is, of course, important to realise that, in some of the instruments cited, the vibrations are from side to side, or *transverse*, rather than longitudinal as was the case for the metal rod. The strings of a piano or harp, for example, vibrate from side to side, but one can still visualise a wave travelling along from one end to the other and being reflected back and forth. This point will be elaborated later.

Unconventional members might be collections of glass bottles, filled with differing amounts of water, hung on strings and hit with a light hammer; pan pipes, which are collections of straws or hollow reeds of different lengths played by blowing across the open end – here the air column is the vibrator and the mechanism of excitation will be discussed in chapter two; sansas or thumb pianos, which are found in many parts of the world and consist of different lengths of springy metal or cane the ends of which are plucked by the thumbs as the instrument is held in both hands. 1.19 is a typical example of the South African sansa from the collection in the Horniman Museum, London.

1.19 *A South African 'sansa' or thumb-piano from the collection in the Horniman Museum*

SOUNDS OF DIFFERENT PITCH FROM ONE VIBRATOR – SMOOTH CHANGES

Sometimes it is possible to change the pitch of the note of a single vibrator in order to play tunes. If, for example, the effective length of the string of a guitar is changed by altering the position at which it is clamped to the neck board by the finger, the length of travel of the wave and hence the pitch of the note can be changed. Alternatively the speed at which the wave travels may be changed by making the string tighter or slacker. In other words we take a single vibrator and alter its shape or its state of tension in order to change the pitch of the note; we shall assign instruments which use this as their main method of pitch changing to family number two. Conventional instruments using this principle would obviously include all the string family – in which the length of the string is altered to play tunes and the tension for basic tuning – though here a small number of different vibrators is usually used to increase the range. The woodwinds belong to this group – in the clarinet, flute, oboe, etc., the particular keys depressed or holes covered by fingers determine the shape of the air column in the pipe, which is the actual vibrator, and hence determine the pitch.

An unconventional member of family number two is the musical saw in which the performer produces changes of pitch by altering the degree of bending. 1.20 shows the method of gripping the saw between the knees and bowing near the mid point of the S-bend.

1.20 *Playing a musical saw; the more violent the curvature near the handle, the higher the pitch provided that the blade always has a double curve like a letter S*

Is the human voice a conventional or unconventional instrument? Either way it is certainly a member of this family. We shall spend much more time considering its complexities later on in the book, but it is worth making two points now. The primary source of the sound we make – whether in speaking or singing – is the vocal chords. These are just a pair of rather crude membranes with a gap between which create a rough basic sound in something like the way small boys and girls produce rudimentary squawks by blowing two blades of grass stretched between their thumbs, or by letting down a balloon with the neck stretched to make a narrow slit from which the air can escape. You can listen to a sound very close to that of the basic sound of your own vocal chords by saying 'Ah' – as when requested by a doctor. But if you sing 'Ah' instead of saying it, you can make the sound move to a higher or lower pitch by more-or-less unconscious adjustment of the muscles of the throat. What you are really doing is to alter the tension of the vocal chords – the higher the tension the higher the pitch. This change of pitch is, of course, also used to introduce variation in speech. (Listen to the two 'Ah' sounds in the question 'Are you in a car?' said in a surprised tone of voice.) But there is a second way in which it can be demonstrated that plain speech is bound up with musical sounds. Pretend that you have 'lost' your voice completely and can only whisper. This means in fact that you have stopped the vibrations of your vocal chords by letting them become very slack and relaxed. Try then to whisper 'Ah' and, without stopping, gradually change to 'Ooh' (as in soup). Do this several times and you will probably be able to sense a change in the pitch of the sound you are producing even though it is only really a rush of air (very like the white noise we mentioned earlier on). If you have some musical training you may be able to judge that the change in pitch is somewhere between a fifth and an octave, depending on your exact choice of vowel sounds. This time you have done it mainly by changing the shape of your mouth cavity in which the air is vibrating. In fact something more complicated than a single change of pitch has occurred, as we shall see in later chapters, but at least we have demonstrated that there are two senses in which the voice belongs to family number two, even when used for speaking.

SOUNDS OF DIFFERENT PITCH FROM ONE VIBRATOR – SUDDEN JUMPS

The saw – or indeed any member of the family discussed in the last chapter – may be used to illustrate another way in which the pitch of a vibrator may change. If one attempts to glide smoothly up the scale on the saw over a wide range, the pitch very often jumps suddenly. If a recorder is blown gently and then more vigorously with the same holes uncovered two quite different notes will result. These are simple examples of what the physicist calls a change of mode of the vibration. It simply means that the *pattern* of vibration is altered and not just its frequency. This system of changing pitch – that of family number three – is used by all the members of the brass family. The post horn or bugle provide good illustrations; there are no means by which the shape or tension of any part can be changed and only one vibrator is provided – but by altering the way in which the waves are set up by the lips the system can be made to vibrate in different modes and hence give a limited sequence of notes. If a suitable instrument is not available, a length of hosepipe or brass pipe or even electrical conduit can be used for demonstration. In the orchestral versions such as the trumpet, trombone, etc., methods of changing the shape (valves, slides, etc.) as well are provided in order to increase the range of notes produced. We shall think in more detail about the mechanisms in later chapters – but now we must consider what happens when the mode changes.

PATTERNS OR MODES OF VIBRATION

A useful way of beginning to understand what we mean by a change of mode is to use a model suggested by – among others – Arthur Benade. Consider a weight fixed at the mid point of a horizontal length of elastic cord. A wooden ball

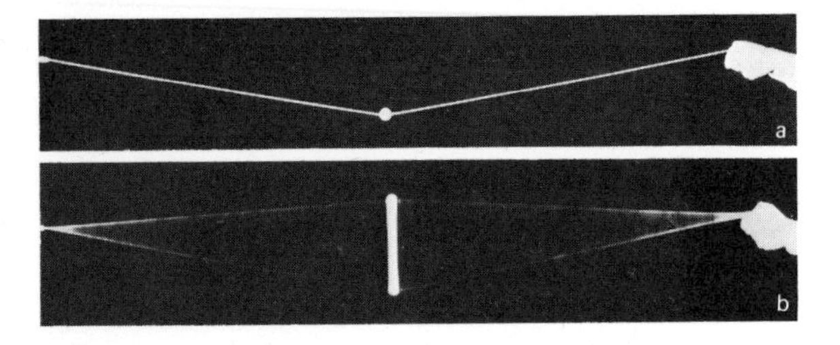

1.21 a and b *Vibration of elastic carrying a single ball at its mid point*

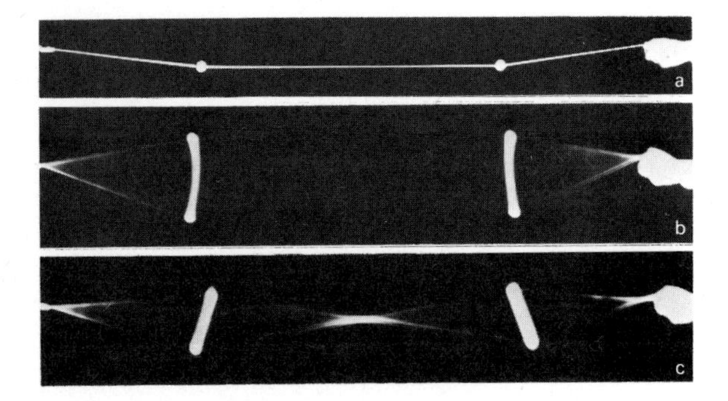

1.22 a–c *Two different patterns or modes of vibration for elastic with balls at two places*

say $1\frac{1}{2}$ inches diameter on a 2-yard length of cord works well (see 1.21). One end of the elastic is fixed and the other moved up and down at various frequencies. It soon becomes clear that there is only one frequency at which the weight will oscillate smoothly and regularly up and down. Now repeat the experiment with two weights placed roughly a quarter and three-quarters of the distance along the elastic (1.22). Now there are two frequencies at which stable oscillations can be produced; in one both weights move together, and in the other one moves up as the other moves down and the second pattern involves oscillations of higher frequency than the first. In the second mode the system behaves just as though it consisted of two separate single-weight systems of half the length and double the frequency. A three-weight system has three modes as shown in Fig. 1.23. This process can be continued indefinitely – ten weights will have ten modes, twenty weights twenty modes, and so on.

Thus if we move over to a continuous rubber rope without any weights – i.e. an infinite sequence of infinitesimal weights – there will be an infinite number of modes. For this system it turns out that the ratios of the frequencies of the successive modes are 1:2:3:4:5:6 . . . etc., to a surprisingly close approximation.

A useful demonstration of this continuous system – which can be considered as a large-scale model of a guitar string – is provided by a length (say about 10 yards) of solid rubber cord. (Cord about 3/16 to 1/4 inch in diameter sold for making gaskets in vacuum systems is ideal.) One end is held stationary by an assistant and the other end is held by the demonstrator with the cord horizontal but not stretched too tightly. The exact tension can quickly be found by experience, but the middle will probably be 6 inches to 1 foot below the ends when the tension is about right. If the rope is plucked downwards about 1 foot at a point about a foot from one end, a wave can clearly be seen travelling up and down the rope taking perhaps half a second or so (depending on the type of rope, tension, length, etc.) to make each traverse. It should be stressed at this point that this is exactly what happens on

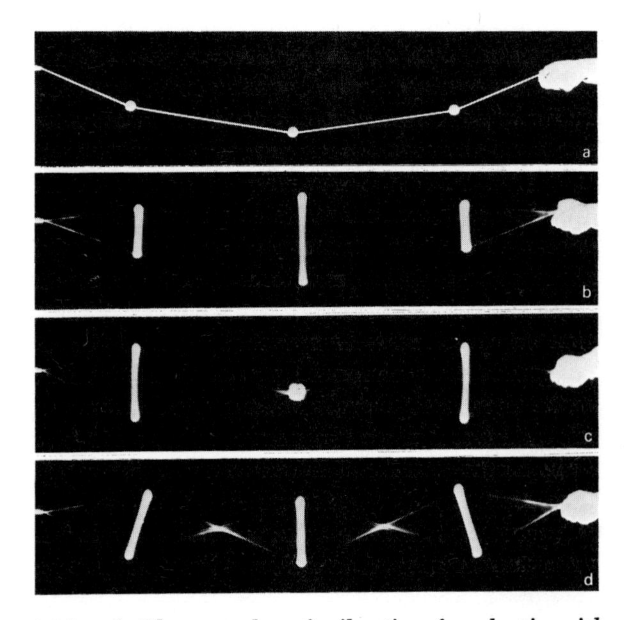

1.23 a–d *Three modes of vibration for elastic with balls at three places*

1.24 a–c *A rubber rope used to demonstrate a wave produced by plucking and releasing at a point near to one end*

a smaller scale, and of course at very much higher speed, when a guitar string is plucked. The fact that a wave travels back and forth is often forgotten because the rapid movement makes it appear that the string is merely moving up and down (1.24). Notice particularly how, when the wave first reaches the right-hand end, it has a shape which is an exact reflection of the shape at the start. The type of motion is, however, very complex and a much simpler form – the fundamental or first mode of the cord considered as an infinite set of particles – can be set up if the demonstrator moves his hand up and down

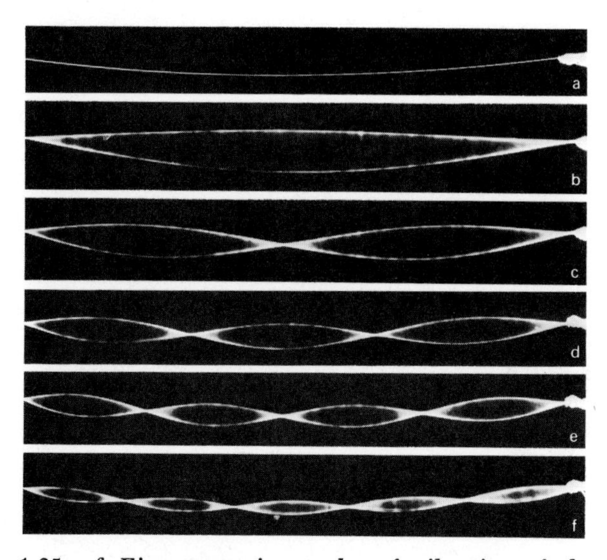

1.25 a–f *Five successive modes of vibration of the same rubber rope produced by raising and lowering the hand on the right more and more rapidly. The rates increase in the ratios 1:2:3:4:5 approximately*

a small amount at exactly the right rate. The first plucking experiment may be used to determine the approximate timing and then, provided that the tension is not changed, it is relatively easy to persuade the string to vibrate with a large amplitude (see 1.25). Higher modes can be produced with a little practice as the rate of vibration is increased successively by factors 2, 3, 4, 5, etc.

Again, although the string appears to be vibrating up and down what is really happening is that two waves are travelling in opposite directions and being reflected back and forth. In the

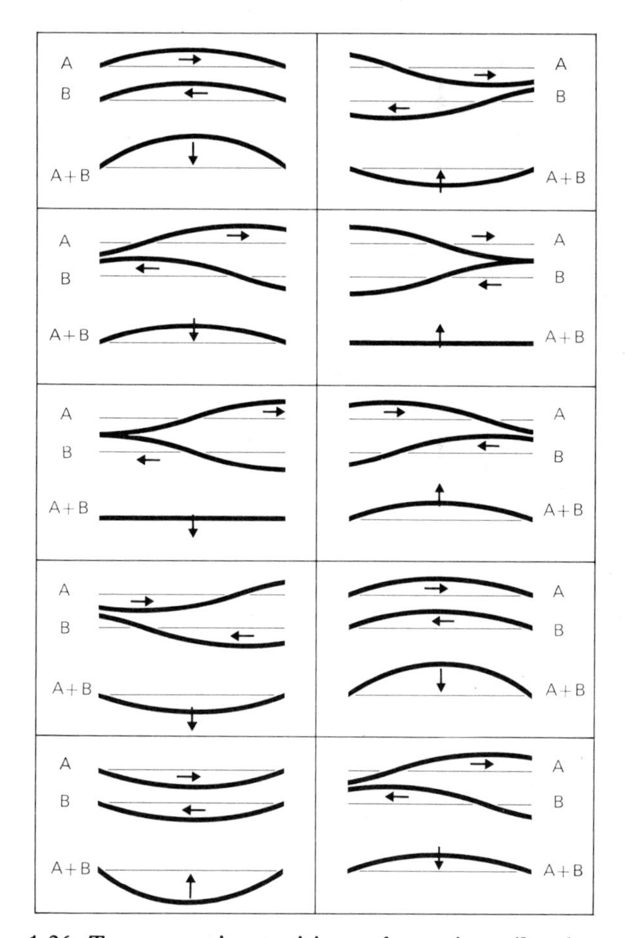

1.26 *Ten successive positions of a string vibrating in its fundamental mode. In each diagram A represents the wave travelling from left to right, B the wave travelling from right to left, and A + B is their sum which gives the resultant shape of the string actually observed. Read down the columns*

fundamental mode the wavelength is twice the length of the string and the succession of positions of the two waves and the way in which they add together is shown in 1.26. If we now make the string twice as long and try to create waves of the same frequency, it is fairly easy to see that the initial displacement will only have time to move half-way along the string before the next wave crest comes along. So, although as drawn in 1.27 the wavelength remains the same as in 1.26, it is now equal to the length of the string whereas before it was twice the length of the string. Now the two halves of the string vibrate symmetrically, one half going down as the other half comes up and vice versa. It is not difficult to see that exactly the same pattern of vibration would occur if we used the *same* length of string as in 1.26 but *doubled* the frequency; the wavelength would now be half the original value in 1.26. For the photographs of 1.25 the string was kept at the same length but the frequency was successively increased.

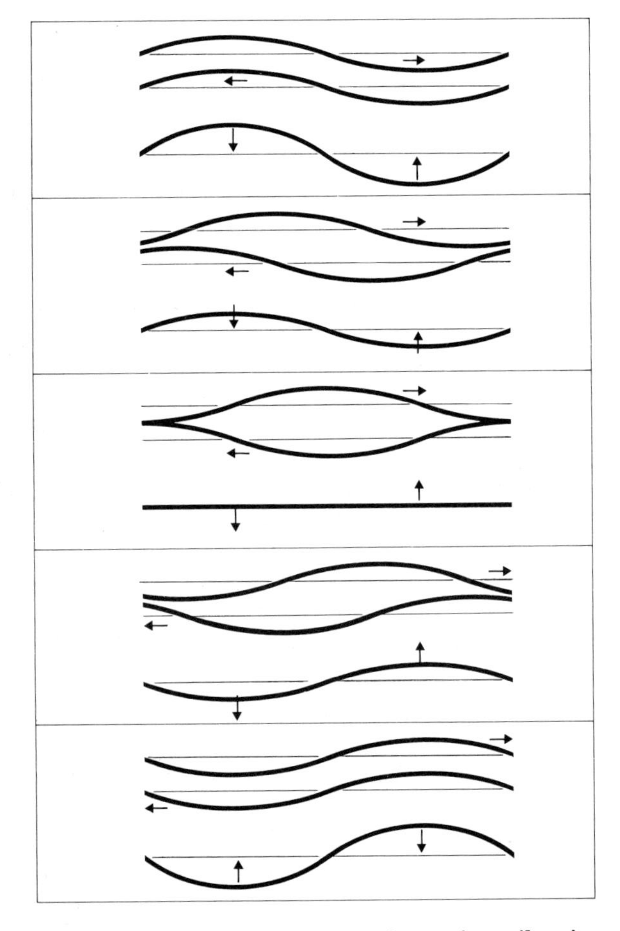

1.27 *Five successive positions of a string vibrating in its second mode. Because the string is twice as long, the frequency will be the same as for 1.26 and the diagram is produced in the same way*

The fascinating point that emerges is that for transverse waves (those in which the displacements are at right angles to the direction of travel of the wave as for a string), and also for longitudinal ones (in which the displacements are along the direction in which the wave is travelling, as for the spring or for sound waves), provided the object is very long and very thin the frequencies of successive modes are roughly in the ratios 1:2:3:4, etc. Modes having these ratios are called harmonic. For systems which are not simple long thin objects the frequencies may not have these simple ratios; the modes are then described as anharmonic.

HARMONICS

The idea of harmonics and harmonic relationships will pop up so often in the book that it is worth spending a moment or two thinking about their musical as well as their physical aspects now. A very convincing demonstration of harmonics may be given with a 'cello – and of course the use of harmonics is a well-known musical device for producing special effects on any stringed instrument. Bow a long note on one of the strings – the highest or A string is probably the best. Then *very lightly* touch the mid point of the string but do not press it down to touch the finger board. With a little practice the octave above the normal note may easily be produced; this is the second mode or second harmonic of the string. If you then slide your finger slowly *away from the bridge* – still touching the string very gently – a sequence of progressively *higher* notes can be produced. At first this seems paradoxical until one realises that the lightest touch on the string prevents the string from vibrating easily at that point and hence it adopts a mode which has a null point (or *node* as the physicist calls it) there. Thus as one moves the finger away from the bridge the string is successively forced to vibrate in the sequence of modes that are easy to see in the demonstration experiment with the rubber rope that is illustrated in 1.28.

There are two points to notice immediately. First the places at which one has to touch the string are successively closer together. This is fairly easy to understand if one looks at the distance between one end and the nearest node in 1.28 b, c, d, e, etc. But secondly the musical intervals become smaller. In fact the musical intervals between the first six harmonics are an octave, a fifth, a fourth, a major third and a minor third. This point may be understood if one plots the successive frequencies on the logarithmic scale introduced earlier and illustrated in 1.14 b. 1.29 a shows the first ten harmonics represented on this kind of scale and one can see immediately how it is that the sensation is of notes being closer together. 1.29 b shows the corresponding musical notation, the frequencies and the musical intervals for ten harmonics on the fundamental A = 55 Hz.

Notice that the intervals decrease progressively and also that the seventh harmonic is given a question mark on both the musical scale and the intervals because for some reason it does not occur in our western musical system. Another point will have been noticed by musically trained readers – that the intervals between the eighth and ninth and between the ninth and tenth harmonics are designated 'major' and 'minor' whole tones with frequency ratios 9:8 and 10:9. But on the piano A to B is the first 'step' (Doh-Ray)

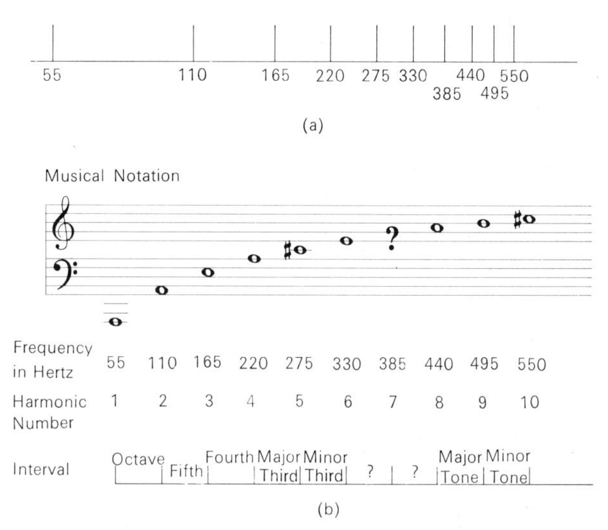

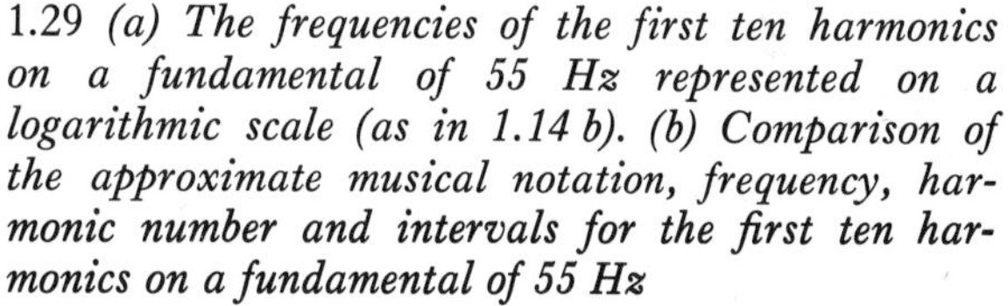

1.29 *(a) The frequencies of the first ten harmonics on a fundamental of 55 Hz represented on a logarithmic scale (as in 1.14 b). (b) Comparison of the approximate musical notation, frequency, harmonic number and intervals for the first ten harmonics on a fundamental of 55 Hz*

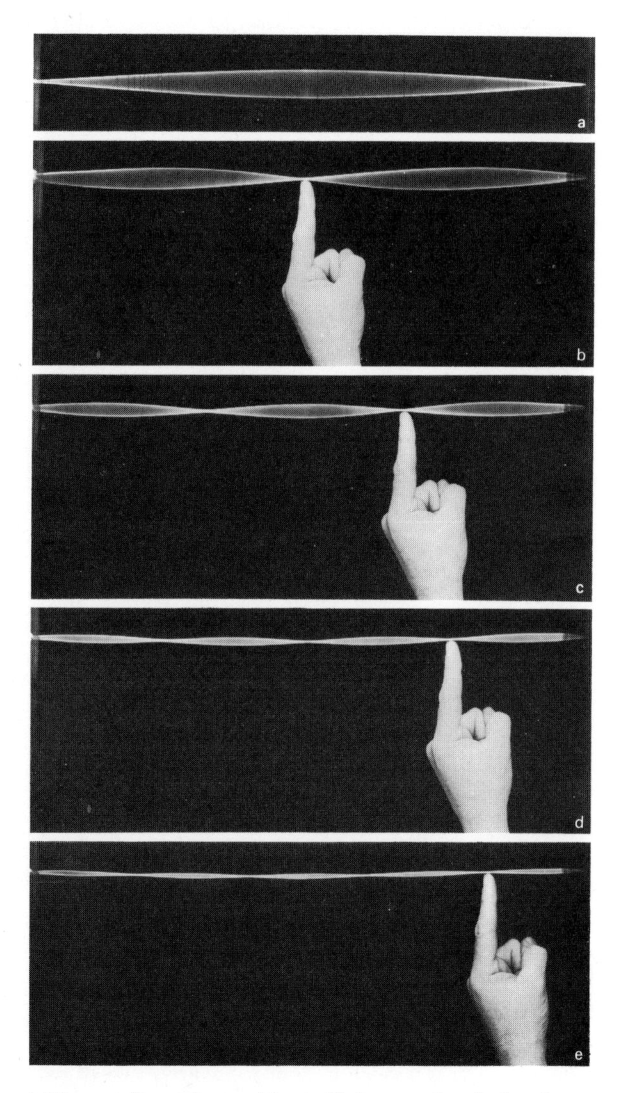

1.28 a–e *Stopping with a light touch of the finger to produce successive harmonics illustrated with rubber cords to exaggerate the effect*

in the scale of A major, and B to C♯ is the first 'step' (again Doh-Ray) in the scale of B major. How can this be? This, of course, is our first meeting with the great problem of the instruments in family number one – those with a separate vibrator for each note (e.g. keyboard instruments) – which does not affect those of families numbers two and three to the same extent. The answer to the problem is that C♯ in the scale of A major is not the same note as C♯ in the scale of B major – in other words the intervals specified are quite correct – but in order to give this precision the number of notes required on a piano would be so enormous as to be impracticable. We therefore compromise and represent them both by the same note by using a special system of tuning which evens out the errors as much as possible. The ideas behind it will be discussed in chapter six when we have studied more about the way in which our ears and brains work. For the moment we must accept it as a practical solution to a tricky problem and remember that Bach wrote his forty-eight preludes and fugues 'für das Wohltemperierte Klavier' precisely to demonstrate how good a solution it is and how successfully one can play acceptably in all keys. If it was good enough for Bach it is probably good enough for a very high proportion even of the most sensitive listeners!

A FURTHER LOOK AT MODES OF VIBRATION

We have said that for long thin vibrators the modes are found to have frequency ratios 1:2:3:4, etc., or in mathematical terms are said to be harmonic. It is, of course, possible to produce mathematical explanations for this, but for the non-mathematical readers perhaps the easiest way to begin to understand it is by means of a diagram. In a vibrating string such as that of 1.25 it is clear that the whole system is symmetrical and, in particular, is the same at both ends. It follows therefore that the two waves travelling in opposite directions, which, as we saw first in 1.26, add up to make the whole pattern, must be producing the same kind of vibration at each end. Consider 1.30; if we start from A the next points where the vibration is the same are B, C, D, etc. Similarly, if we start at A′ the next points are B′, C′, D′, etc. This tells us that if we can fit in a pattern of length AB with a particular string or pipe so that the vibration is the same at each end, then we shall also be able to fit in AC if we divided the wavelength by 2, AD if we divided the wavelength by 3, etc. Since the speed of the waves is the same, a wave of half the length will take half as long to pass by and hence the *frequency* will be doubled. We can see therefore that the frequency ratios of possible patterns will be 1:2:3:4:5, etc. Now it does not necessarily follow that all these patterns are possible in *all* thin systems. One of the earlier members of family number one about which we talked was a brass rod clamped at its centre. Clamping at the centre means that no movement is possible there and hence only those patterns which not only have the vibrations the same at both ends but also have zero vibration at the middle will fit. 1.31 shows how the slinky spring can be used to show

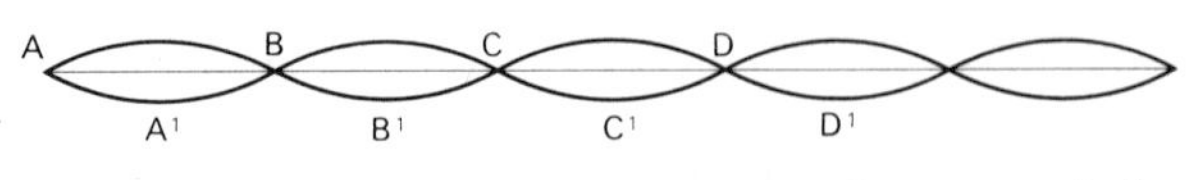

1.30 *An aid to the discussion in the text of the sequence of harmonics in a free, symmetrical long thin vibrator.*

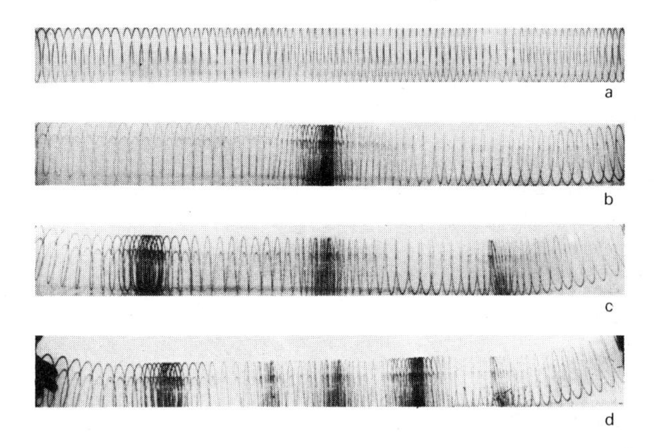

1.31 a–d *Three modes of vibration of a spring with symmetrical movement at each end but always with a fixed point at the centre. The frequency ratios or harmonic numbers are 1:3:5*

the kind of vibration pattern that occurs in the rods and it is easy to see that there is only one zero point (the centre) in 1.31 b, but there are three in 1.31 c and five in 1.31 d. The frequencies are in the ratios 1:3:5. In other words, every other harmonic is missing in this example. It turns out that exactly the same pattern of harmonics occurs in a tube which is open at one end and closed at the other – in fact this behaves exactly like half of the system shown in 1.31 and the zero vibration at the centre corresponds to the closed end of a pipe. Once more we can use the argument about fitting the waves in, but this time we must look for places where the waves are *different* at each end. In 1.32 the simplest pattern that would fit with maximum at one end and zero at the other would be AB.

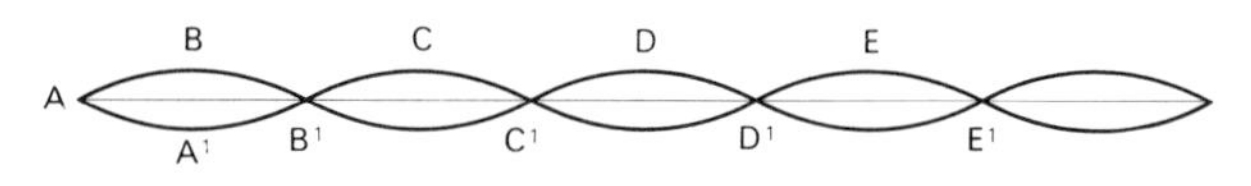

1.32 *An aid to the discussion in the text of the sequence of harmonics in an unsymmetrical long thin vibrator or in a symmetrical one with restraints such as a fixed point at the middle*

The next simplest could be AC if we divide the wavelength by three and then AD if we divide by five. It is thus possible to see how in the unsymmetrical case with a pipe closed at one end and open at the other, or for a rod forced to be stationary at its centre and hence behaving like two unsymmetrical vibrators put together, the sequence of modes has every other harmonic missing.

The kind of question that a physicist would immediately want to ask is 'Can we have only these two extremes – what would happen if the system were somewhere between complete symmetry and complete asymmetry?' The answer is much more complicated and much more like what really happens in musical instruments as we shall see in chapter four. The ratios are no longer exactly whole numbers and the extent to which they depart depends on the exact way in which the pattern at each end differs.

Another very important factor is the uniformity of the system. 1.33 shows a set of three

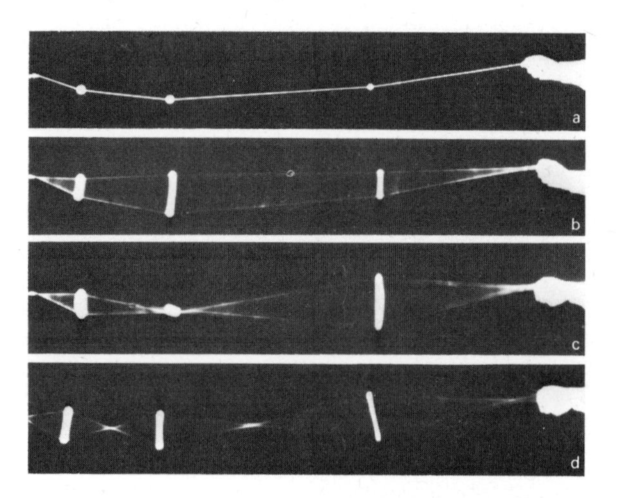

1.33 a–d *Three modes of vibration of elastic with three unequal balls distributed non-uniformly; however, the frequency ratios are no longer simple*

weights on a string unequally spaced and unequal in size: it is clear that the vibration pattern is a complicated one and the wavelength *is different at different positions.* How can this be? The frequency of vibration must be the same all along the string, otherwise one could not have stability at all, but the speed at which the waves travel now varies – it is in fact slower at the heavy end where the weights are bigger as one might expect. If the wave gets slower at the heavy end, then a crest will not travel as far in a given time and so the wavelength gets shorter; in general the frequency ratios of successive modes are no longer simple. This is something like the kind of behaviour we meet in tapered or non-uniform pipes and in a later chapter we shall have to come back to study the point in more detail. For the moment, however, it merely serves to remind us that we must not jump to the conclusion that there are nice simple explanations waiting to be found for all the phenomena we shall be studying; in most cases the full story is very complicated indeed.

HOW FAR HAVE WE PROGRESSED ON THE JOURNEY?

We should now pause to consider how much progress we have really made so far in making and

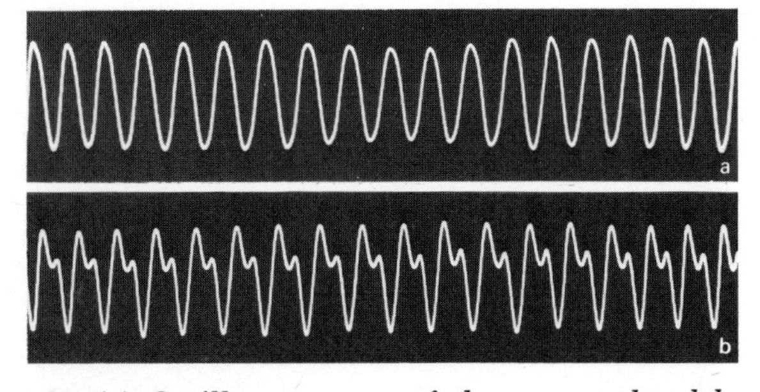

1.34 *(a) Oscilloscope trace of the note produced by stroking a brass rod with a resined glove (approx. 1700 Hz). (b) Oscilloscope trace of a 'cello producing the same note as does the rod in (a)*

measuring waves and in the journey of the waves to our ear.

We have seen how scientists look at the graphical form of the pressure variations, how the pitch we perceive is roughly related to the frequency we can measure in Hertz, and how the loudness we perceive is related to the amplitude or size of the pressure variations in the waves. We have considered some simple ways of setting up the periodic waves in the air that correspond to simple musical tones and have seen already how the three principal families of melodic instruments in the western orchestra arise naturally out of simple scientific principles – though clearly from a historical point of view the science came afterwards. We have also just begun to catch a glimpse of some of the complexities which will face us very soon when we begin to move away from the very simplest kinds of vibrations. In other words we have not gone very far at all. But some foundations have been prepared and we know how to create simple musical sound waves in the air; the next question that we must consider is just how the waves produced by real musical instruments differ from those produced by simple vibrations like the musical saw or the brass rods. 1.34 is a comparison of the wave produced by one of the brass rods of 1.18 with a note of the same pitch produced high on the first string of a 'cello. The rod produces a very simple kind of wave which suggests that it may be vibrating in a simple mode. But for the 'cello the pattern of vibration is much more complex – although the frequency at which the pattern itself repeats is the same as that for the rod. What could the extra 'kinks' imply? They clearly represent changes in the waves occurring *faster* than those for the rod; that is, they must represent higher frequencies of some sort. Is it possible that the 'cello string is vibrating in more than one mode at once? That would explain the presence of higher frequencies and might begin to explain the quality differences. It turns out, however, that this represents only a small part of the answer and, in any case, we shall have to explain how a vibrator can be made to vibrate in more than one mode at once. We shall therefore shelve this question for a little while and devote ourselves in the next chapter to the more fundamental problem of how any kind of vibration is started up and see how the waves in the air develop from small beginnings.

On the facing page:
Concert champêtre, from an old engraving

Three ways of beginning notes: Percussion, string and wind instruments, all in Daumier's painting: Parade de Saltimbanques

FROM SMALL BEGINNINGS

THE START OF A NOTE

If you have ever listened to a piper starting to play the Highland bagpipes you will recall that, as the bag is filled with air and then struck to set the reeds vibrating, for the first few seconds the sound is quite different from that produced when the music proper begins. In this example, part of the delay arises from the time taken for the pressure in the bag to build up to a steady level, and hence the starting time is much longer and the difference in quality is much more noticeable than for most other instruments. It turns out, however, that although we may not be specifically conscious of it, the way in which a note starts for all instruments is of great importance in determining our ultimate reaction when we hear it. As will become increasingly obvious, the process of hearing is a very complicated one and, in order to recognise a particular instrument or voice, we need to piece together a great many different clues. The first split second of any note or sound gives us the first of these clues and can be said to trigger off the processes of sifting, sorting and comparing that go on in the brain. The more characteristic and sharply defined the beginning, the more rapidly we are able to identify the sound. Later we shall consider many ways of studying the small beginnings and their consequences but, to demonstrate immediately that we are not exaggerating the importance of the start, consider the words 'pa' and 'ma'. Their meaning is significantly different and the brain must rapidly recognise the difference. 2.1 shows oscilloscope traces of each, spoken by the same person with a deliberate attempt to achieve identical vowel quality and length. Each sound lasts half a second (500 milliseconds or ms). But you can easily see that for about three-quarters of the time the sounds are identical; the difference occurs in less than a quarter of the whole sound. The two different consonants are just two different ways of starting up the same sound.

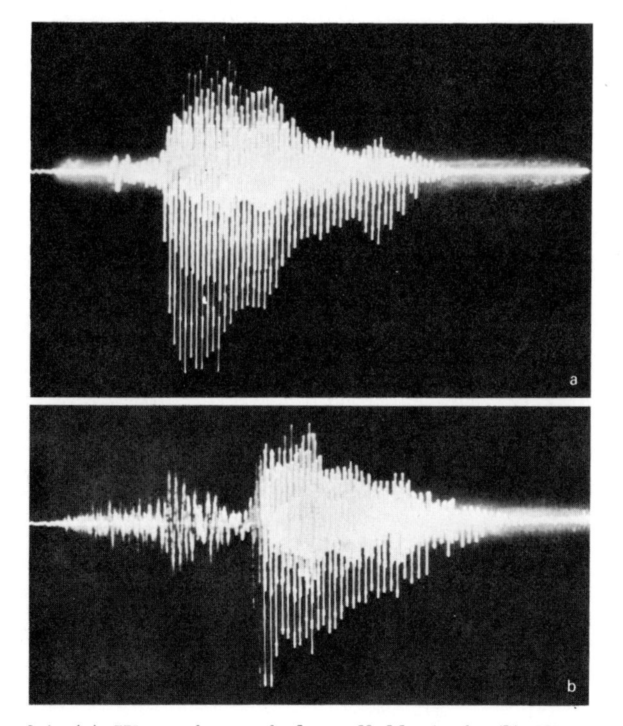

2.1 *(a) Wave form of the syllable 'pa'. (b) Wave form of the syllable 'ma' using the same speaker and conscious attempt to make the vowel part the same*

Perhaps an even more striking illustration – which is really a preview of a later section – is obtained if we look at the oscilloscope trace of the word 'far', again spoken by the same person with a conscious attempt to achieve the same vowel quality as before and making the vowel part the same length – though the 'f' takes longer than the 'p'. Its trace is shown in 2.2. Now compare with 2.1 a which looks exactly like the trace for 'far' except that it starts very suddenly towards the end of the 'f'. In other words the sound

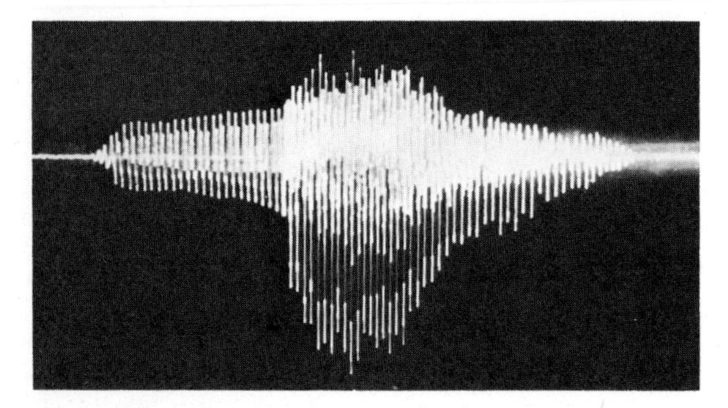

2.2 *Wave form of the syllable 'fa' again with the same vowel quality*

'p' is almost exactly the same as 'f' but is much shorter and starts rather more suddenly. If the word 'far' is recorded on tape and the beginning of the tape is literally cut off and replaced by blank tape it is possible to match the trace for 'pa'; even better, if you can try the experiment yourself with a tape-recorder you will find that it *sounds* exactly like 'pa'. Does this convince you of the importance of the start?

EVEN A SIMPLE VIBRATOR TAKES TIME TO SETTLE DOWN

Before we study real sound generators it will help us to form a mental picture of the processes that go on if we think of a very simple model of a vibrating system – a pendulum.

Imagine a bob about the size of a billiard ball hanging on a length of string about one metre long. If we allow it to swing freely this pendulum will move back and forth and the bob will trace an arc one metre in radius (2.3 a) and it will take about two seconds to make one complete swing, that is to move from one end to the other and back to the start again.

How do we start a pendulum swinging? The simplest way is to pull it to one side, hold it still with the string taut (2.3 b) and release it. On the face of it we might expect that the bob would start to move along its arc immediately from A, accelerating towards the lowest point B and then slowing down and coming to rest at C, starting to accelerate back to B again and so on. In other words, we might expect the motion to begin instantly and to exhibit no variations at all (except of course that the arc would gradually become shorter as the motion died away, though with such a system this would happen only after a very large number of swings).

But we have jumped to quite a few conclusions and if we examined the motion very closely and carefully we should find that we have much over-simplified what actually happens. In order to visualise the effect more readily we must change our mental picture slightly. Suppose that we replace the string with very 'stretchy' elastic; now when the bob is released it tends to extend the elastic and so moves well below the arc (2.3 c). But then the elastic contracts again and pulls the bob back towards the arc and beyond it so that it follows a path which is quite complicated (2.3 d). You can imagine that it may be many swings before the bob adopts the smooth undisturbed arc as in 2.3 a.

Now we come to the more difficult point to

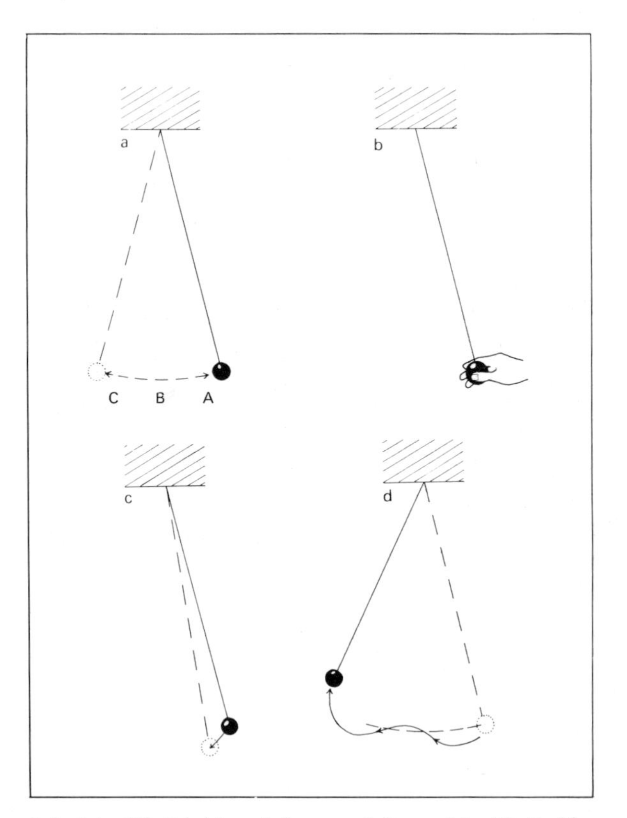

2.3 *(a), (b) Rigid rod for pendulum. (c), (d) Rubber cord for pendulum*

visualise; the complicated motion of 2.3 d happens with *every pendulum* even if the support is thick string or a solid steel rod. The only difference is that, whereas with the elastic the variation from the arc may be measured in centimetres (hundredths of a metre), with the steel rod it may only be measurable in microns (millionths of a metre). What we are really saying is that it is virtually impossible to start a system vibrating instantly in the way that it will continue; it will always start with variations or perturbations of a greater or lesser extent and for a longer or shorter time (depending on how small are the variations we can measure or detect).

It should be clear from this discussion that, once again, the all-important factor is time, and whether or not, in listening to a note produced by a mechanical vibrator, we are aware of an effect at the beginning depends on how long it takes for the particular note to settle down compared with the time taken by our ears and brains to start working. As we shall see later, the all-important times may be as short as 20 milliseconds – as for a consonant such as 'T' – or as long as 500 or more milliseconds for the build-up of the note of a large organ pipe. It should also be clear from our very simple pendulum model that the *way* in which we make the vibration start also has a profound effect on the time taken to settle down. Thus if we start with the pendulum at rest and strike it a sharp blow (think of the elastic pendulum for the clearest mental picture) the subsequent behaviour will depend on the strength of the blow and on its precise direction (i.e. whether horizontal or at an angle). Clearly also it would start in a very different way if a gentle jet of air were directed momentarily at the bob while hanging in its undisturbed position. As always analogies should not be pushed too far; the pendulum picture is only given to help you to see that the way vibrations start is important. In this chapter then we shall need to look carefully at all the ways of starting up vibrations, and in order to do this we shall need to devise techniques for studying the first half-second or so of a note in very great detail.

Before plunging in, however, there is a complication which occurs in nearly all real instruments that must be considered, although it is concerned not only with the small beginnings but also with the growth of the note which we shall study in the next chapter.

IN MOST REAL INSTRUMENTS THERE IS AN ADDED COMPLICATION

The complication is that very few instruments consist of a single vibrator. A string vibrating on its own can communicate its motion only to a very small amount of air and so produces a very quiet sound indeed; as soon as it is coupled to a sounding board (as in the piano or harp) or to a hollow body (as in fiddles or guitars) much larger vibrations can be set up in the air, leading to much louder sounds. In the woodwind and brass instruments the primary vibrators are the reed, the air striking the edge of a flute or whistle, or the lips of the player. Their vibrations are coupled to those of the air in the pipe and so again we have at least two parts to the system. The aspect of the multiple system that concerns us in this chapter is that it leads to extra complications during the starting period. Let us take the piano as an example. It is clear that when the hammer strikes the string, which is quite a violent interaction, the initial vibration is likely to be different from that a little later on when the string is vibrating freely. But from the instant the string is struck it must share its energy of vibration with the sounding board and, since this is a fairly massive piece of wood on a strong metal frame, it will certainly take time and delay the growth of the note. Perhaps an even easier example to picture mentally is that of a tuning fork. Struck on its own it produces a very quiet note; stood on a table, however, the note can be heard all over the room. This time the fork is the primary vibrator and it has to communicate some of its energy to the table and, by making the whole table top vibrate, create waves of larger amplitude in the air and hence louder sounds in the ear. We do not, of course, get something for nothing, and the compensation is that the fork gives up its energy much more rapidly so that the note dies away very quickly when the fork is stood on the table. What happens as the

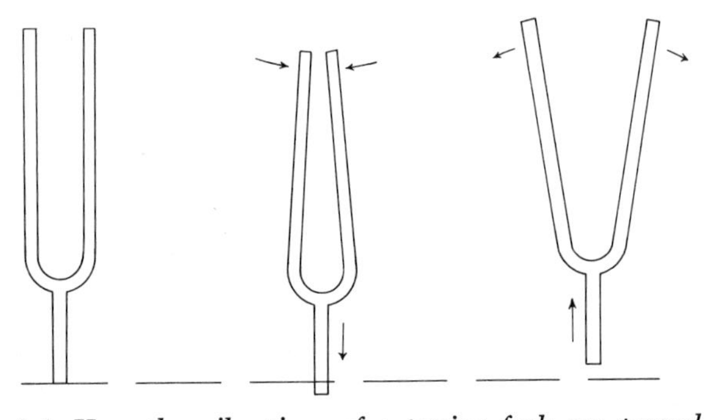

2.4 *How the vibrations of a tuning fork are passed on to the table*

fork is brought into contact with the table? The prongs of the fork (2.4) vibrate in and out and make the stem vibrate along its length; as the stem touches the table top the vibrations spread out as waves through the thickness of the wood and along its surface, and – without going into any details at all – one can see that the whole table top cannot instantly spring into vibration but time will be taken for it to start moving. If you listen carefully as the fork is slowly lowered on to the table, the first tenth of a second or so can be heard to have a quality of sound distinctly different from that of the steady note eventually produced.

The process can be likened to an argument in which the table is at first reluctant to play a part in the sound production but eventually succumbs to the persuasiveness of the fork. We shall come back to this picture of an argument on several future occasions and will find that the result is not always so one-sided; sometimes the result is more of a compromise. A good example would be the oboe, in which the reed on its own produces a squeak or squawk depending on how it is blown; when fitted to the instrument the air column is persuaded to vibrate by the reed, but the length of the column – fixed by the keys and the player's fingers – has strong 'views' about the final pitch of the note and reacts back on the reed until eventually a steady note is set up by the partnership. Whatever the outcome of the argument the technical term given to the initial part of any note that is significantly different from the steady part is the 'starting transient'; 'starting' because it happens at the beginning and 'transient' because it soon disappears and is replaced by a steady note.

So far we have talked round some of the complications involved in starting up a note but have not really grappled with any practical details. One thing remains to be done before we can proceed, and that is to discuss some special tricks and techniques of measurement and observation that will be needed in our study.

SPECIAL TECHNIQUES FOR STUDYING THE FIRST SPLIT SECOND OF A NOTE

There are three main techniques that are valuable and – perhaps not surprisingly in view of the stress already put on this aspect – all are really concerned with the manipulation of time. The first is to vary the speed of the trace on the oscilloscope. 2.5 shows the trace of a single note – middle C – played on a small harpsichord in which the string is plucked mechanically and the vibrations then die away freely. In (a) the spot travels from one side to the other in 1/2 second and we see that there are just over 13 peaks corresponding to a frequency of 262 Hz. In (b) the spot travels across in 1/4 second and, though it is much more difficult to see the shape of each individual crest – or even to count the 66 crests – we can see the relationship of one to another in size and can see, for example, that this trace corresponds to part of the note that is dying away. In (c) the spot travels across in 2 seconds; the individual waves can no longer be distinguished, the picture of the changes that occur both as the note builds up and as it dies away is much more complete. Thus by varying the speed of the trace we can control the magnification of our 'time-microscope', which is what our oscilloscope really is. In studying sounds that are very short-lived it is useful to use an oscilloscope with what is called a 'long-persistence' screen. This is one in which the trace goes on glowing after the spot has passed to give time for the trace to be studied. 2.5 d shows the first 1/100 second of the same note and this could only be seen with a long-persistence screen – or of course with some kind of photographic recording device. Appendix IV, in which oscilloscope

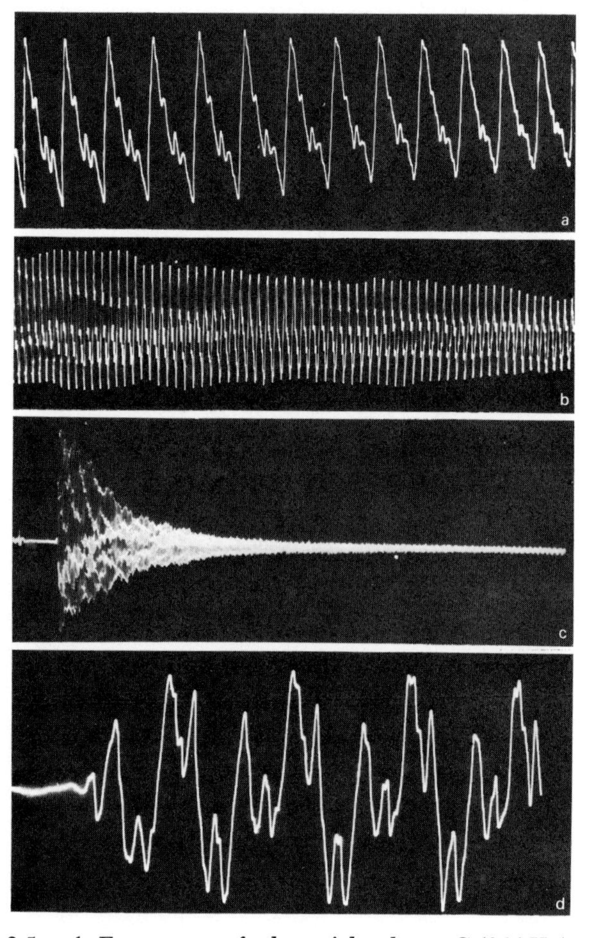

2.5 a–d *Four traces of a harpsichord note C (264 Hz): (a) lasts 1/20 second; (b) lasts 1/4 second; (c) lasts 2 seconds; (d) lasts 1/100 second*

traces are compared with a musical score on various time scales, may be helpful in getting this technique clear in one's mind. Even with a long-persistence screen and the possibility of varying the time of each sweep one still has to be pretty skilful to capture precisely the bit of the signal one wishes to study. This is where the second special technique becomes useful. The sound is recorded on an endless loop of tape so that, as the tape goes round and round, the sound is repeated over and over again as often as one wishes to hear it or display it on the screen. The length of the loop and the speed of the tape determine the repetition rate. 2.6 shows a tape-recorder which has been fitted with a home-made jockey pulley which is held in place by the heavy cylinder which sits on a rubber pad. This system can accommodate loops of a sufficient range of length for most studies of single notes or of single consonants. If the tape-recorder is for stereo, the sound studied can be recorded on one track and a very short burst of high-pitched sound (e.g. about 2000 Hz for 1/100 second) can be recorded on the second track. This second signal is used to 'trigger' the oscilloscope – that is, to start the spot of light on its trip across the screen. By varying the position on the tape of the trigger signal in relation to the signal being studied it is possible to display any desired part of the wave on the screen. A final refinement that is possible is to introduce, by means of an electronic device which we shall not describe, a variable time delay between the trigger signal and the moment at which the spot starts to travel across the screen. With this device we can vary the section of the wave studied without the need to re-record the trigger signal at different positions on the tape. Thus, although the start of a note is over in a fraction of a second we have a means of capturing it in time and can study it with our time-microscope at leisure.

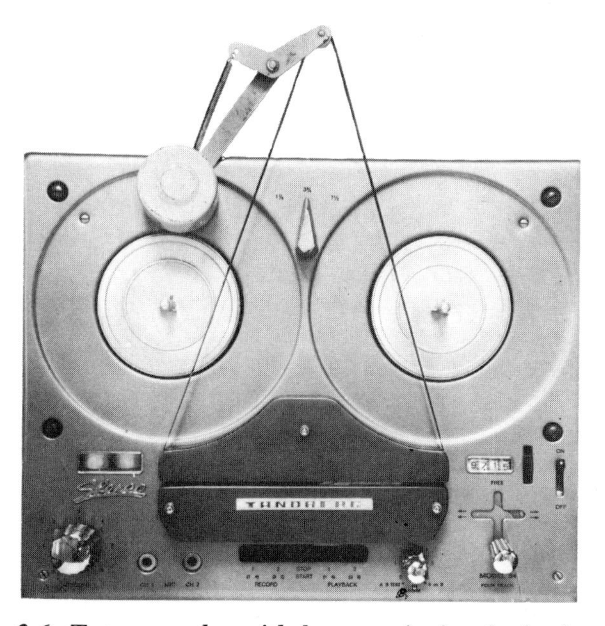

2.6 Tape-recorder with loop-tensioning device in position

The third technique that proves to be immensely useful is perhaps the simplest. It again involves juggling with time – now by varying the speed of play-back on the tape-recorder. Even domestic instruments often have three speeds ($7\frac{1}{2}$ inches per second, $3\frac{3}{4}$ i.p.s. and $1\frac{7}{8}$ i.p.s.), but professional machines are available with wider ranges, or can easily be adapted to give wider ranges. It thus becomes possible to record at say $7\frac{1}{2}$ i.p.s. and play back at $1\frac{7}{8}$ i.p.s. – a quarter of the speed. If this resulting sound is re-recorded at $7\frac{1}{2}$ i.p.s. and again replayed at $1\frac{7}{8}$ i.p.s. we have a total reduction of sixteen to one in the speed. The result, of course, is that the frequency of all the notes goes down to one-sixteenth of their usual value and hence the pitch goes down four octaves. But of much greater importance is that the all-important starting section now lasts sixteen times longer and we can actually hear the changes taking place during starting as well as see them on the screen. A particularly convincing use of this is in studying the initiation of the note of a practice bagpipe chanter. Slowed down to one sixteenth of the speed the steady note sounds rather like a motor cycle pop-popping – but at the beginning we can distinctly hear four or five pops which occur quite irregularly before the steady rhythm begins. (This example is discussed more fully in the section on blowing a reed and is illustrated in 2.23.)

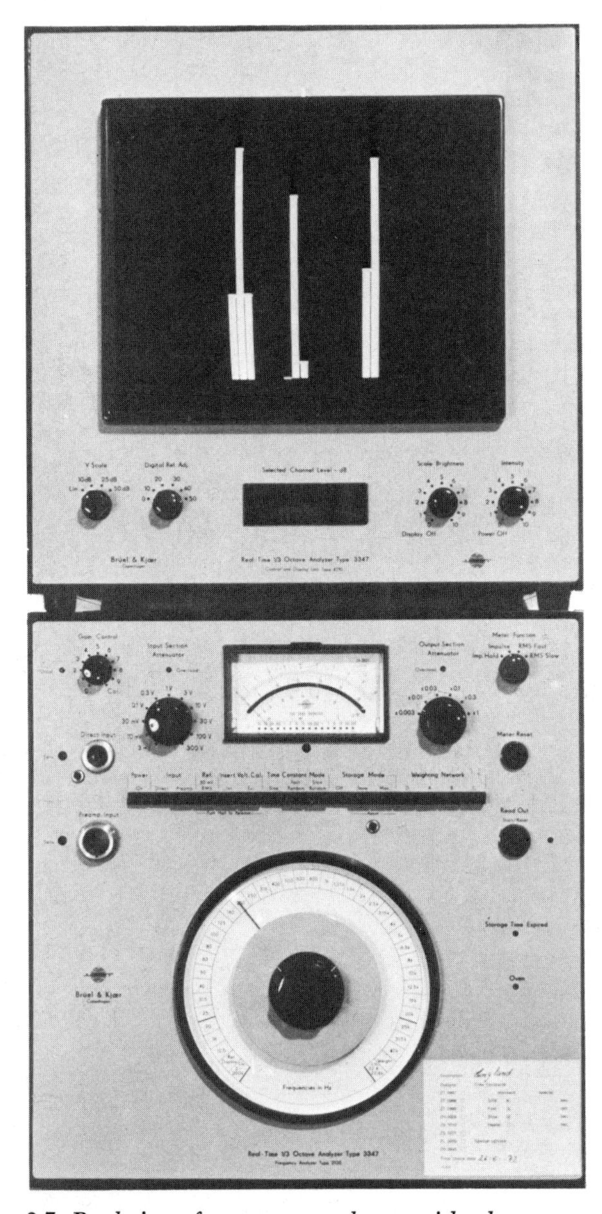

2.7 *Real-time frequency analyser with three pure tones fed in simultaneously at 100 Hz, 404 Hz and 3000 Hz: the nearest analyser bands are 100 Hz, 400 Hz and 3150 Hz*

One further instrument which is invaluable throughout all sections of this work is a frequency analyser. We shall say much more about this in the next two chapters, but for the moment it will be sufficient to describe it as a device which tells us the relative proportions of high and low frequencies that are present in a mixture. 2.7 shows the real-time analyser used for most of the subsequent illustrations; it is particularly useful because it displays the analysis instantly even while the sound is changing and enables one to study variations with time. Three pure tones of frequencies, 100 Hz, 404 Hz and 3000 Hz, are being fed in simultaneously, each with a different amplitude. The analyser has a total of thirty vertical strips on the screen each corresponding to a band of frequencies one-third of an octave wide so that every third strip corresponds to a frequency jump of one octave. The lowest tone falls exactly in the centre of the corresponding frequency bands of the analyser and this is confirmed by the symmetry of the small strips on either side. The middle note is 1 per cent higher in pitch than the nearest analyser band and this shows up in the asymmetry of the small strips on either side. The highest note is 5 per cent lower in pitch than the nearest band and so the pattern is even more asymmetric and the upper side strip has disappeared. (A quick calculation shows that the frequency ratio between each successive band and the next lower one is 1.26 : 1, since $1.26 \times 1.26 \times 1.26 = 2.0$, the ratio for 1 octave.) Now we have our instruments and techniques and can start looking at the ways in which notes are brought into being on real instruments.

A CLOSE LOOK AT WAYS OF INITIATING NOTES AND AT THEIR CONSEQUENCES

(a) Striking – for example, bells and drums

It might be thought that the closest approach to an instant start would be that obtained by striking an object. This is not so, however, and the very violence of the act of striking, combined with the inertia or reluctance of the object to start moving, leads to marked changes during the initiation period; they give a characteristic quality which helps us to recognise instruments in this category even when competing with a great deal of other sound. The bells in Tchaikovsky's '1812' overture, for example, though not particularly loud, can nevertheless be heard and recognised against the full orchestra.

Let us try to think through what happens when a bell is struck – and to begin with we will consider the tubular bells of the kind used in the orchestra or in door chimes. The essential feature is a length of hollow tube, hanging from one end, struck with a hammer, usually fairly close to the suspension point. As the hammer strikes the bell there is a sudden violent compression of the metal at the point of contact; there is also violent compression of air between the hammer and the metal and it rushes out as the space between decreases. The air then rushes back, the hammer may bounce and hence there are a few large pulses produced in the air which travel out at the beginning of the wave and which, if we listen carefully, are heard as a non-musical thud. The compression of the metal also travels as a wave in all directions, but in particular it travels to the free end of the tube and is then reflected back. If the metal is 'springy', for example, tempered steel or hard brass, the compression may travel back and forth along the tube many thousands of times before it eventually is damped out; gradually, however, the resonance of the tube asserts itself and it begins to vibrate strongly in one of its modes and so a musical note merges out of the initial thud; its pitch is determined mainly by the time taken for the pulse to travel up and down. If the metal is soft, for example, lead, the internal friction may be so great that the wave is damped out almost immediately and all we hear is the initial thud. This is an over-simplification, and waves will not only travel up and down the tube but in all sorts of other directions as well and hence the note produced may not be a simple one. The tube has many modes of vibration just as the rope did in the last chapter and will probably vibrate in several at once, and furthermore, because of the way the vibration is started some may involve vibrations round the circumference as well as along the length.

In a church bell – though rather more complicated – the essential principles are the same. The clapper or tongue strikes the sound bow – the thickened part of the rim of the bell – and the energy communicated to the bell in this initial crash is then gradually channelled into various modes of vibration of the bell. Normally we are not particularly conscious of the crash, but if a recording, say of Big Ben striking one o'clock, is played at a very slow speed the emergence of the musical tone out of the initial crash is much more apparent. In a BBC Radio 4 programme called 'The Swinging Giants' such a recording was played backwards so that the musical part was heard first, gradually rising to a crescendo and then becoming the crash; it is surprising how much more obvious the crash is in the reversed recording.

In chapter one we talked about modes of vibration, but only for long thin vibrators for which we found that the frequency ratios were harmonic, that is they have the whole number values

1, 2, 3, 4, 5, 6, etc. In chapter three we shall return again to the idea of modes but will extend it to two- and three-dimensional objects. It should be fairly obvious now, however, that as soon as we depart from long thin objects there are so many directions in which the waves can travel that it is highly probable that the complete sets of waves can be fitted in (following the idea illustrated in 1.30 and 1.32), in many more ways than would lead to simple harmonic ratios. It is not surprising, therefore, to find that the church bell with its complex shape has a great many modes and that they are anharmonic. Since we are at present only concerned with initiation we need not pursue this particular aspect further except to point out that striking a vibrator does tend to initiate a larger number of modes of vibration than do most other methods of starting notes. Later when we discuss the modes of flat plates we shall take up this point again.

The kind of discussion followed for bells can be applied equally well to all the percussion instruments. Some, of course, have less well-defined musical pitches largely because the sound is of such short duration that we do not have enough time to identify it positively. If compared with another sound of different pitch, however, we rapidly become aware of the difference: hollow coconut shells used to imitate the 'clip-clop' of horses' hooves would be a good example. Tunable orchestral drums – the tympani – are complex systems because not only do the modes of vibration of the stretched skin play a part, but also those of the three-dimensional air cavity and, to a lesser extent, those of the metal shell. As far as initiation is concerned, however, the principles are much the same as those for bells.

Perhaps the instrument which best demonstrates aurally the points we have been discussing is the large gong or tam-tam. The initial crash is obvious – then the deep note emerges – but because it is associated with a vibration of the very large area of the whole gong this is damped out fairly quickly and we hear the higher frequencies persisting for much longer. The quality of the note produced is thus continually changing; but the continuation of the change really belongs to the next chapter.

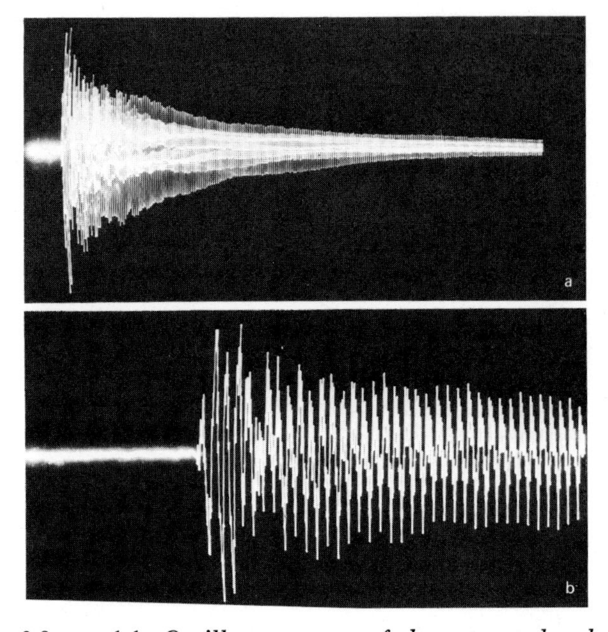

2.8 a and b *Oscilloscope trace of the note produced by plucking the top (E, 330 Hz) string of a Spanish guitar – the other five strings being damped: (a) whole note, (b) about the first 1/10 second*

(b) Plucking – for example, guitars and harps

Plucking a string turns out, in fact, to be one of the most rapid methods of initiating a musical note. 2.8 shows the oscilloscope trace of the note produced by plucking the top string of a Spanish guitar (E, 330 Hz); in (a) the whole note is displayed with a slow scan and in (b) about the first 100 ms are shown. It clearly does not take long for the note to start and become established. It is well known, however, that guitar and harp players can produce a great variety of different tonal qualities from the same instrument. It would appear that variations in the start are not solely responsible, as we have already seen that the note from a plucked string is quickly established. We can find a clue by using the frequency analyser. 2.9 a is the oscilloscope trace of the note of the top string of the guitar as in 2.8 recorded just after the vibrations have steadied down after the transient and before the long decay has gone very far. The time scale is much faster and the whole trace now represents only 30 ms. 2.9 b is the frequency analysis of this

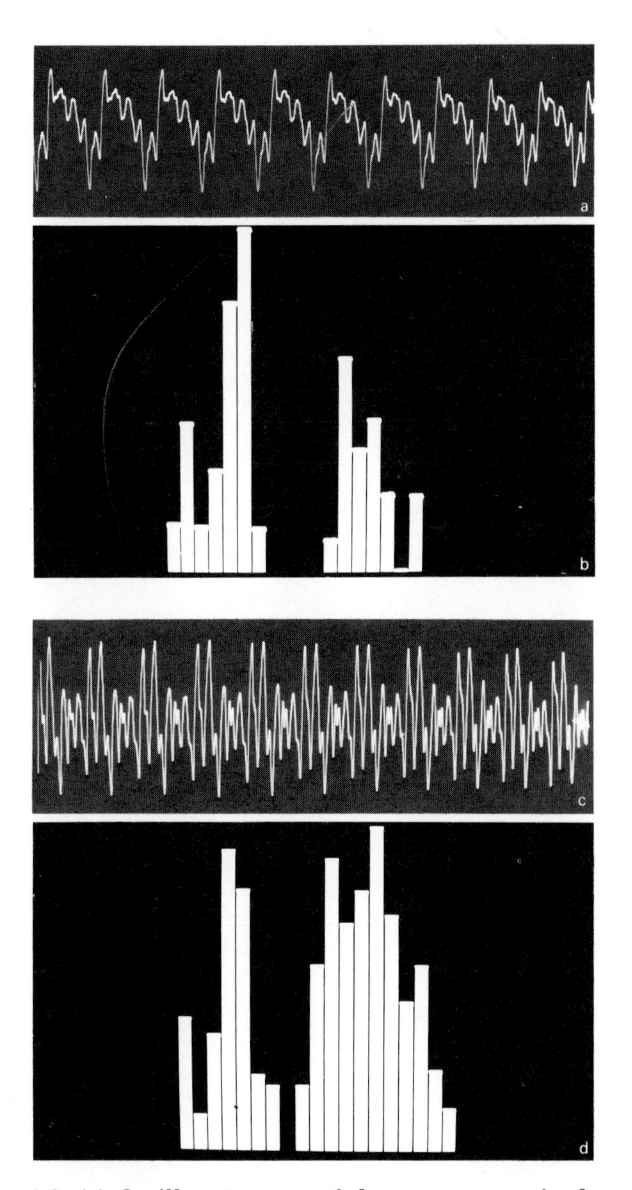

2.9 *(a) Oscilloscope trace of the same note as in the last figure recorded just after the maximum amplitude is reached and lasting only about 30 ms. (b) Frequency analysis of the same note. (c) As for (a) but plucked very close to the bridge. (d) Frequency analysis of note of (c)*

note. The next three pairs of photographs give traces and analyses done in the same way, on the same string of the same guitar, but for (c) and (d) the string is plucked very close to the end of the string nearest the bridge; in 2.10 a and b the string is plucked at the normal point, but with the ball of the thumb used sideways to give as broad a contact with the string as possible; in 2.10 c and d the string is plucked at the normal point, but a pin is used as a plectrum. It is easy to see the variations in wave shape and corresponding frequency analyses that can arise. It seems, therefore, that the nearer the point of plucking moves to the end of the string the higher the proportion of upper harmonics, and also that

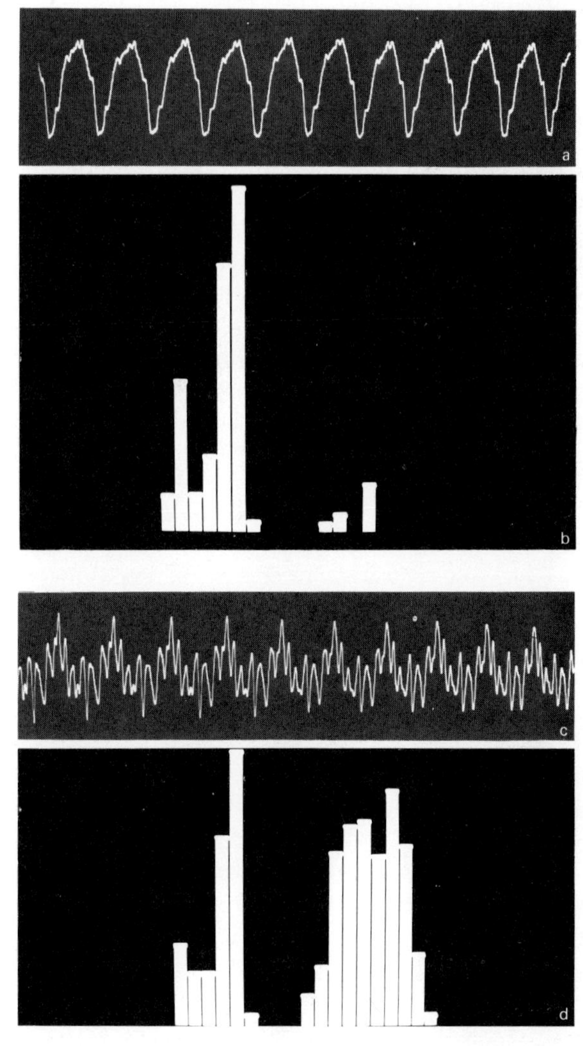

2.10 *(a) As for 2.9 a but plucked with the broad ball of the thumb. (b) Frequency analysis of note as (a). (c) As for 2.9 a but using a pin-point as a plectrum. (d) Frequency analysis of note (c)*

narrower and harder objects used to pull the string to one side also tend to increase the proportion of upper harmonics. That this is very reasonable and fits in with some of the ideas developed in chapter one can be seen if, once again, we make use of mental pictures, or better still of some simple experiments.

Perhaps this would also be a suitable point at which to comment on the dangers of relying too much on mental experiments. This was the Aristotelian method and the great danger is that one can so easily visualise the wrong result! A classical example of this that has nothing to do with music is that of two soap bubbles blown on opposite ends of the same tube. The question is 'What happens if one is larger than the other?' Most people instinctively feel that the larger one will shrink and blow up the small one until they become equal in size. In fact the larger one grows and the smaller one shrinks almost to nothing! The moral is that attempts to visualise are excellent in teaching and exposition, but should as often as possible be checked against real experiments, and must never under any circumstances be used alone as a basis for research. In this book we are using them as a means of explaining principles already established by experiment and trying to convince the reader of their reasonableness! Ideally every reader should try to carry out at least some of the experiments in person, though clearly some of them cannot be done outside an acoustics laboratory.

Now to return to the method of setting a string in vibration. In chapter one we talked about models for the modes of vibration of a string and in 1.24 we saw what happens when a large-scale model of a guitar string is plucked. Look again at 1.25 a; can you think of a way in which the string could be set in motion instantly but in this mode only? Suppose we cut a piece of wood to match the profile of the extreme position of the string, used this to displace the string and then, by moving the template sideways, let the string slip off the edge (see 2.11 a). It is fairly clear that just this one mode will be produced and what we have really done is to use the broadest possible plectrum. If we wished to excite only the third mode by plucking – again

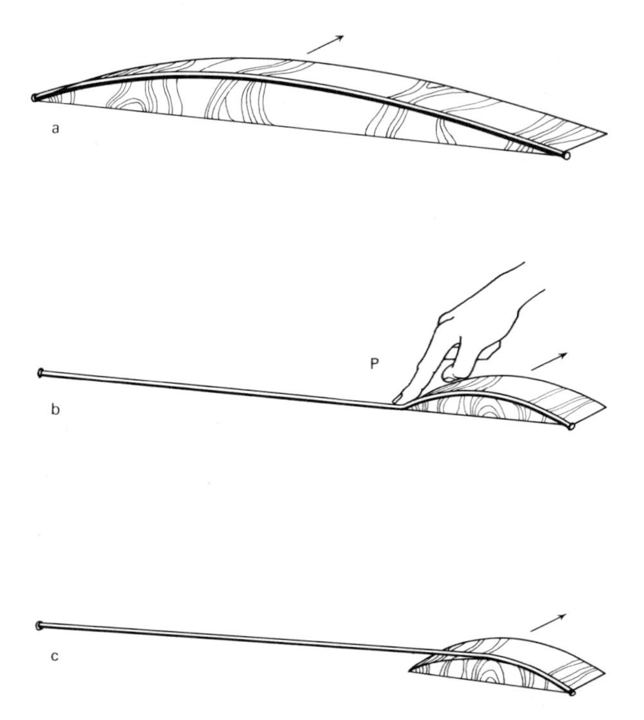

2.11 *(a) Not very practical method of setting a string vibrating in one mode only by sliding the wooden template in the direction of the arrow. (b) Even less likely method of setting up a higher mode. (c) Designed to stimulate thought about what happens when a broad plectrum is used*

consider 1.25 c – it might be possible if we made the template correspond to one of the vibrating sections; in displacing the string we should have to place a restraint at the first node or point of no displacement (point P in 2.11 b). If we did not place the contact at P it is fairly easy to see that we could equally well excite any of the modes 1 to 3 with this template since, once more looking at 1.25, the shape of the string close to the end is very similar in all three. It is unlikely, however, that we should start up any higher modes, because these would have nodes nearer to the fixed end and our template of 2.11 b forces the string to move at all points between P and the end.

Is the explanation beginning to make sense now? One can see how both the breadth of the plectrum and the point of plucking affect the number and relative proportions of the modes established and, at least in principle, one can see that the picture agrees with what we found in the guitar experiments. This is perhaps a rather crude approach and mathematicians have a very much more elegant way of dealing with the problem, but that is beyond the scope of this book; it will be hinted at towards the end of this chapter.

It is important to realise, however, that when we talk about vibrations in several modes at once the individual modes cannot be seen – even if one uses the rubber rope of 1.24 and 1.25. What one sees is the result of adding all the separate modes together in the way which is elaborated later. 1.24 shows the complex resultant for a large number of harmonics since a narrow plucking instrument (relative to the length of the string) was used at a point close to the end. 2.12 shows a sequence of photographs of the same rope taken by the light of an electronic flash; this is a lamp which emits very short flashes of light at times that can be accurately controlled. The timings have been chosen so that the position of the rope can be seen at four or five instants between the time of release and the moment when the wave is reflected from the far end. In the picture the wave starts out travelling from left to right. Notice that, except for the short section parallel to the position of the rope at the left-hand end when the motion is started, the other two sections of the rope remain stationary. If we fixed our attention on a particular point of the rope, say one-eighth of the distance from the right-hand end, it would remain stationary below the undisturbed line for quite a while; then when the wave arrived it would flip rapidly over to the upper position, stay there for a short time till the wave was reflected back and then flip back to the lower position and stay there all the time until the wave returned again. A point at the middle of the rope would stay in the upper position for roughly half a cycle and in the lower position for half a cycle. This may seem an overemphasis on detail, but it is important to get the point clear as the result is totally different from that observed for bowed strings.

Most of the points that have been made in relation to guitar strings apply equally well to the harp and to instruments in which the plucking is done mechanically as in the harpsichord. We shall return to the specific features associated with these instruments in chapter four. There are special tricks and techniques of tone production in all plucked instruments, but it would be out of place in a book of this size and scope to go into further detail; the essential scientific

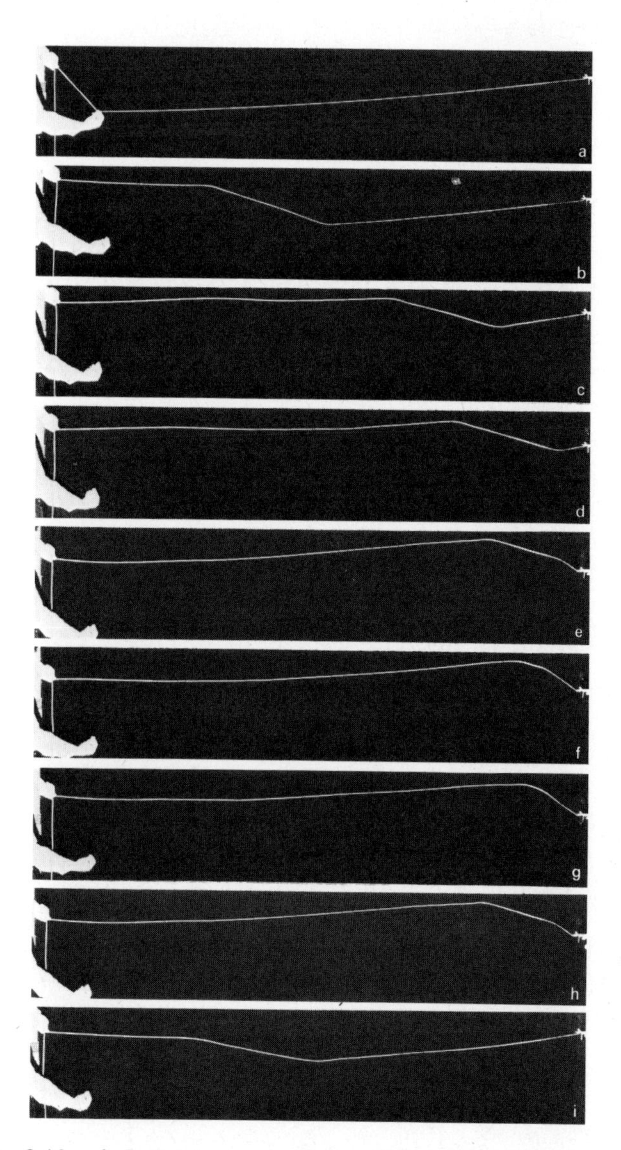

2.12 a–i *Instantaneous flash photographs of the wave travelling along a rubber rope displaced as at (a) and then released. In (a)–(f) it is travelling from left to right and in (g)–(i) it is returning from right to left*

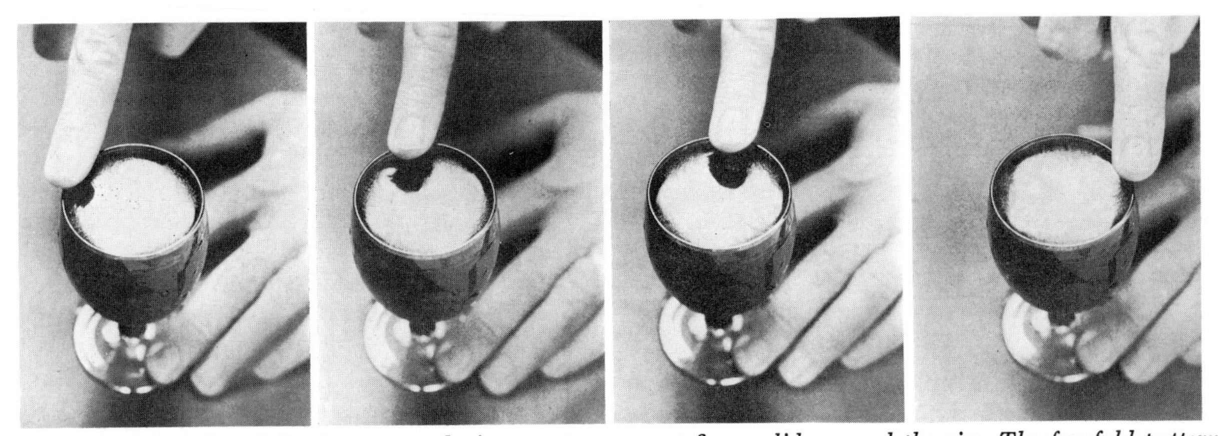

2.13 a–d *Wine glass full of water producing a note as a wet finger slides round the rim. The fourfold pattern of vibrations which disturbs the water surface rotates with the fingers*

principles have been described and it should be possible to extend these basic ideas to enable more elaborate methods of playing to be understood.

(c) Bowing – for example, fiddles of all shapes and sizes

In chapter one we talked about the method of producing sounds from brass rods by stroking with resined gloves and the phenomenon was described as 'stick-slip' motion. We must now consider this type of motion in much greater detail as it contributes essential features of the tone quality produced by all the members of the string family of the orchestra.

The second example of stick-slip motion is a very familiar one. Everyone must at some time or other have produced a penetrating note by running a moistened finger round the rim of a wine glass. The secret is to have the glass very clean and both glass and finger thoroughly wet with detergent-free water. The water in some areas of the country is less effective than others and a trace of water-soluble gum – such as gum tragacanth – is sometimes helpful. The motion is again of the stick-slip type between the finger and the glass, but there is an additional interest here in that the mode of vibration actually rotates! The mode usually produced has four null points equally spaced round the rim and four points of violent motion. As the finger moves round, so the whole mode pattern moves round. 2.13 shows a glass in vibration – it has been filled with water and the outside has been blackened to make the surface pattern stand out more clearly. If a set of four weights is attached (with epoxy-resin adhesive) symmetrically to the glass the rotation can be stopped (2.14). The result then is to produce two notes – rather like the horn of a police car – as the finger goes round; the higher note occurs when the weights are at nodal points and the lower when the weights are at anti-nodes or points of maximum vibration. Again the impor-

2.14 *Wine glass with weights attached: when this is stroked the nodes do* not *rotate, but the glass emits one of two notes depending on whether the finger is above one of the weights, or not*

tant point at the moment is that the notes build up gradually.

A mechanised version of the musical glasses was invented by Benjamin Franklin and is known as the Armonica. Glass bowls of varying sizes are mounted on a rotating spindle and set in vibration by the moistened fingers of the player. Mozart wrote an *Adagio* for the instrument and the fact that it is adagio underlines the fact that the notes take a considerable time to emerge. The author had the splendid opportunity of attempting to play one of these instruments (2.15), which is in the Music Collection at the Horniman Museum, and can vouch for the fact that not only does the vibration take a considerable time to start, but that the process of playing is somewhat nerve-racking because of the delicacy of the old glass and the considerable pressure required to produce a note! 2.16 a is an oscilloscope trace of a note for this Armonica, lasting about 1 second, and the very slow initiation and rather shorter decay are both very clear; 2.16 b is a portion lasting only 1/25 second and shows the relative purity of the note.

We should now move on to consider the commonest and most important example, musically speaking, of stick-slip motion – the bowing of stretched strings. Bowing has been studied scientifically by a great many researchers, including Helmholtz (1877), Sir C. V. Raman (1918) and many recent workers who have made use of the advances in electronic technology. Although there is still room for debate on some of the finer points of detail the broad principles are fairly well agreed. The operation depends on the frictional properties of certain surfaces which lead to the resistance being less when the two surfaces are in relative motion than when stationary. This is a fairly common phenomenon: place a block of wood on another flat board of similar wood and slowly tilt it; an angle can be found at which the block will just slide down the slope. Reduce this angle very slightly and the block will stay in position if placed carefully on the board; give the block a slight push, however, and it will move down the slope and continue to move to the bottom. The dynamic friction is less than the static friction.

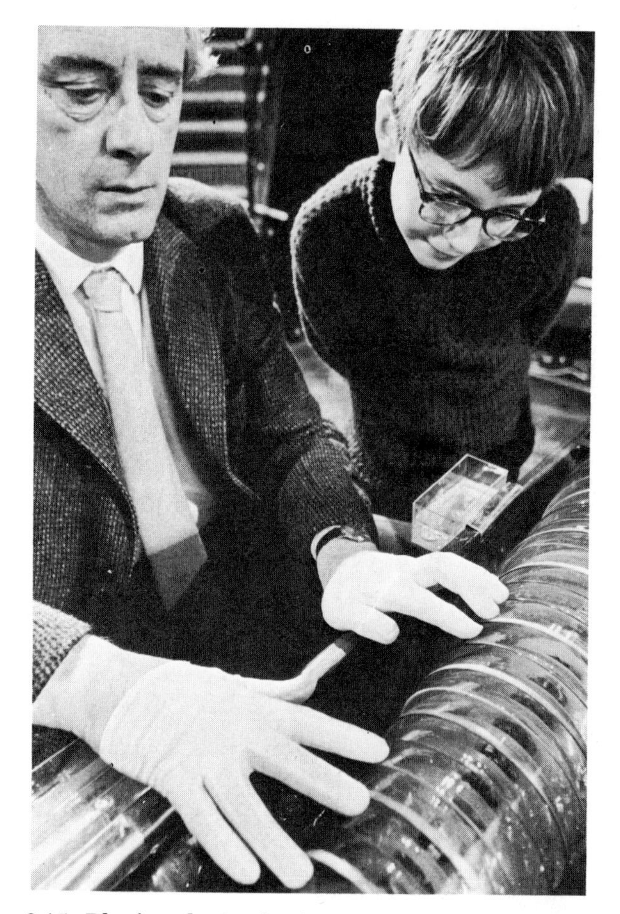

2.15 *Playing the Benjamin Franklin Armonica from the Horniman Museum*

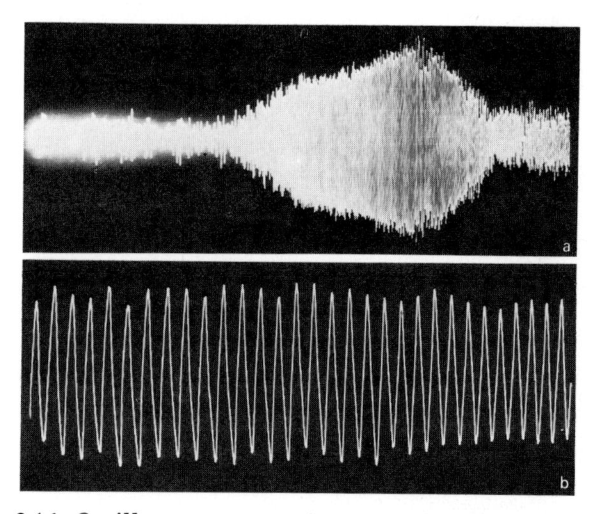

2.16 *Oscilloscope traces of note produced by glass Armonica: (a) whole note, (b) about 1/25 second*

Now consider a violin bow placed across a string and moved steadily to one side. At the moment when the bow is placed on the string the two are not in relative motion and the high static friction is invoked; the string therefore sticks to the bow and begins to be displaced. Gradually as the string is drawn to one side the restoring force increases until it just exceeds the static frictional force and the string slips. Immediately the much lower dynamic frictional force replaces the higher static one and the string goes on slipping past the bow until it not only reaches its undisplaced position but its inertia carries it on. Now the restoring forces of the elasticity of the string come into play again and it gradually slows down, comes to rest and starts to accelerate in the forward direction again. All this time, of course, the bow has been moving at a constant speed in the forward direction. Shortly after starting the move in the forward direction again the string will find itself travelling momentarily at the same speed as the bow – i.e. there will be no relative motion, the two will 'stick' and static friction again takes over and the whole cycle is repeated.

This is perhaps a slightly over-simplified view but is certainly as close as we can expect to approach the real behaviour without plunging into highly complex mathematics. The vital questions are 'How does this behaviour differ from that of plucked strings?' and 'What are the features of bowing that affect the tone quality?'

In the last section we saw that when a string is plucked it flips over from one position to another, then flips back; the time spent in each position may differ, but the 'speed of the flip' is the same in both directions. If you now think of the description of bowing in the last paragraph you will immediately realise the difference. In bowing the string moves slowly over from one position to the other (at a speed determined largely by the bowing speed) and then flips back very rapidly (at a speed determined by the string). This behaviour is quite easy to confirm experimentally by viewing, say, a 'cello string bowed steadily, with a stroboscope, which effectively slows down the motion by any desired time factor, or alternatively by ultra-slow-motion cine photography. It is thus not surprising to find that the proportions of harmonics present in bowed strings differ somewhat from those of plucked strings.

The properties of the string determine the times at which the changes of direction occur at either extreme of the vibration and hence, of course, determine the pitch of the note; the speed of movement of the bow determines how far to one side the string moves and hence the loudness of the sound.

The player has considerable control of the quality just as in the case of plucked strings. He can control the position on the string at which the bow is applied and this affects the proportion of higher-frequency components in much the same way as for a plucked string; he can vary both the pressure and speed of bowing. It turns out in fact that both these affect the quality as well as the loudness; the speed comes in because it obviously alters the speed of the forward movement of the string and hence the ratio of the time spent moving in each direction within the total vibration time fixed by the string; the pressure comes in because it affects the frictional forces involved and can have considerable effect on the damping. Bowing with great pressure produces dull tone, but on the other hand too light pressure leads to uncertainties and may produce only higher harmonics with little or no fundamental.

It is obvious, therefore, that the great skill of violinists involves instant sensitivity to the precise pressure and velocity required and this in turn means that the bow is of as great importance as the fiddle. Much more investigation of the acoustics of fiddles than of bows has been done, but some recent papers in the newsletter of the Catgut Acoustical Society of America consider the problem of bow design. Although it is easy to see how critical the bow is as the link between the player and the instrument, it is by no means easy to specify what the requirements are in scientific terms.

In an earlier section I made the point that physicists always react against the notion of two extreme patterns of behaviour with no intermediate stages. It will not be surprising, therefore,

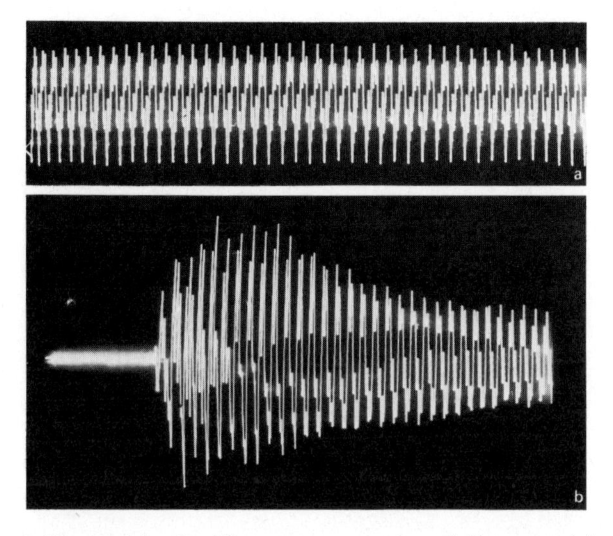
2.17 a and b *Oscilloscope traces of a violin note: (a) bowed, (b) plucked*

to find that the distinction between 'bowed' and 'plucked' behaviour is not absolute. The sliding of the ball of the thumb or finger just before leaving the string has been used to modify the incisive start characteristic of plucking; the so-called 'martellato' and 'spiccato' types of bowing in which the bow is brought into sudden and violent contact with the string to give a very vigorous 'attack' contain an element of plucking or striking to begin with. In general, however, the most characteristic feature of normal bowing is that the note builds up gradually and tends to have a relatively stronger fundamental than for a plucked string. 2.17 and 2.18 summarise these points; 2.17 a is the oscilloscope trace of a bowed open 'A' string on a violin and 2.17 b is for the same string plucked with the ball of the first finger at approximately the same position as that at which the bow was applied in 2.17 a. 2.18 a and b are the corresponding frequency analyses, in both cases taken just after the point of maximum amplitude had been reached.

(d) Blowing on to an edge – for example, whistles and flutes

The wind whistling in the rigging of a ship, the humming of telegraph wires, the fluttering of a flag in a stiff breeze are all closely linked scientifically to the way in which the sound is initiated in a penny whistle, in a flute or in a 32-foot-long diapason organ pipe! The link is the production of eddies or vortices whenever an object moves through a fluid (or a fluid moves past an object – which is the same thing to a physicist – only the relative motion is significant).

2.19 shows the result of a basic experiment which is quite easy to perform. A large, shallow dish contains milk to a depth of about half an inch (dried skimmed milk is fine!) and has graphite dust (lead-pencil scrapings) scattered on the surface. A cylindrical object about 1 inch in diameter such as a section of broom stick or

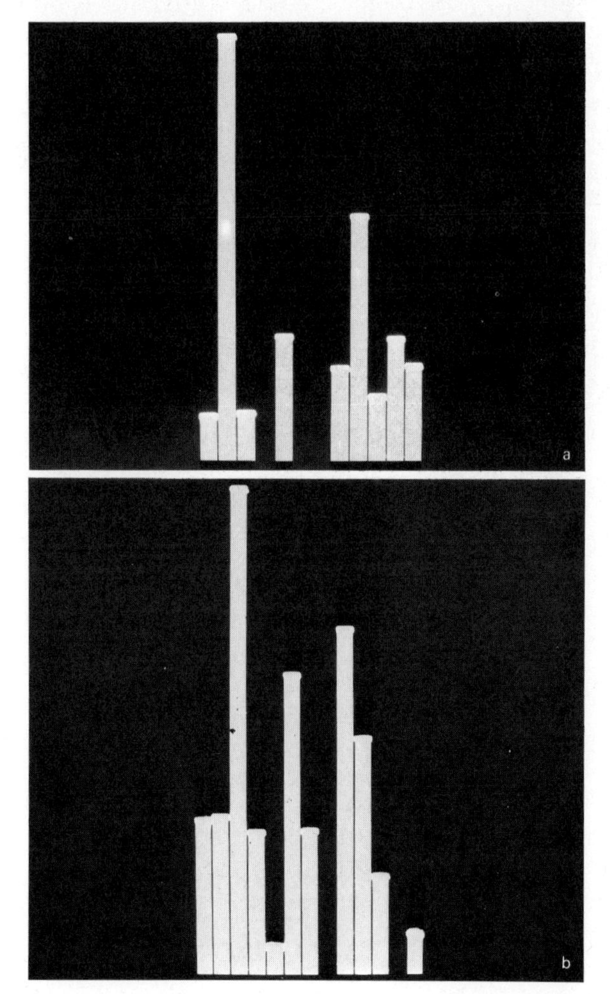
2.18 a and b *Frequency analyses of the notes of 2.17 a and b*

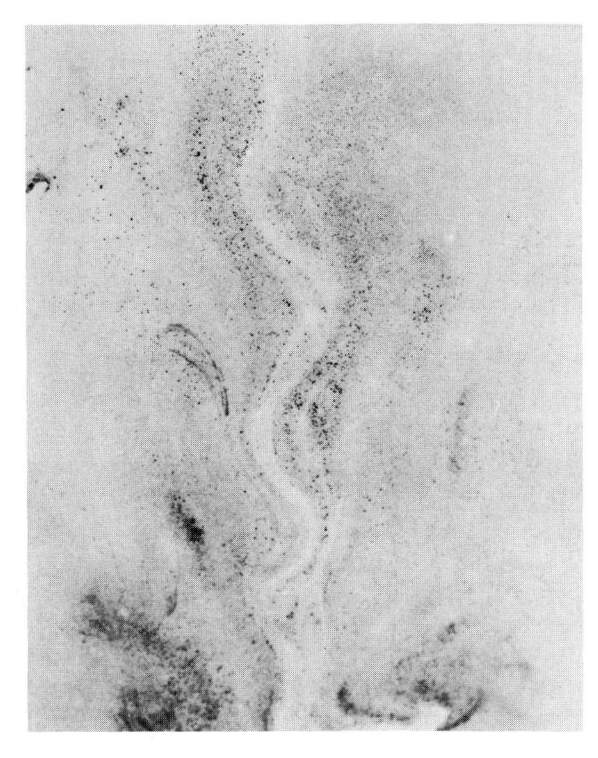

2.19 *Vortices on the surface of milk dusted with graphite*

a large test-tube is held vertically with its lower end immersed in the milk and touching the bottom of the dish. With a little practice one can discover the optimum speed at which to draw it in a straight line across the dish in order to produce eddies which are shown up clearly by the graphite on the surface. The important point to notice is that the eddies occur first at one side and then the other.

Let us now think back to the examples mentioned at the beginning of this section. The rate at which the eddies are produced depends, among other things, on the relative velocity. When the wind blows at varying speeds through the ropes and past the masts and spars of a ship the rate of production of eddies will vary and we hear the characteristic rising and falling pitches. There is, of course, a reaction on the object and it will tend to move away from each eddy in turn (pass a walking stick rapidly through the water in a pond and the reaction making the stick wobble from side to side can easily be experienced). When the wind blows past telegraph wires that are stretched fairly tightly it may happen that the sideways reaction coincides with one of the modes of vibration of the wire and a note is produced which can be heard very clearly if you place your ear to the base of a wooden telegraph pole. In this form the sounds produced are sometimes called Aeolian Tones. The Victorians were fond of a device known as an Aeolian Harp, which consisted of a set of wires stretched on a sounding board which could be placed in the gap at the bottom of an open sash window. As the wind velocity changed, different frequencies were produced and the various strings responded to produce rather weird ethereal chords which seem to grow out of nowhere. Finally the flag 'waves' because the eddies produced by the flag pole or wire on which it is supported pass alternately down each side and provide a very graphic illustration of a vortex train.

How are these vortices used in instruments? The cylindrical object is usually replaced by a wedge-shaped edge which, in effect, separates the two sets of vortices, one set on each side of the wedge. In crude terms then, when a whistle, flute, or organ pipe is blown a set of vortices is produced and alternate ones travel up the inside of the pipe and up the outside. The effect is thus to provide a regular succession of puffs of air travelling along the inside of the pipe and, if the rate at which they are produced happens to coincide with one of the modes of vibration of the air in the pipe, a loud musical note can be produced.

This is a gross over-simplification and all kinds of factors such as the angle of the wedge, the thickness of its edge, its distance from the slit through which the air emerges, whether it is exactly opposite the centre line of the slit, the velocity of the air and others should be taken into account. These variables are particularly important in organ pipes and give the organ builder one of the means of controlling the 'voicing' of the pipes, which really means the way in which the vortex production is matched to the particular modes of the pipe to give the desired quality of sound.

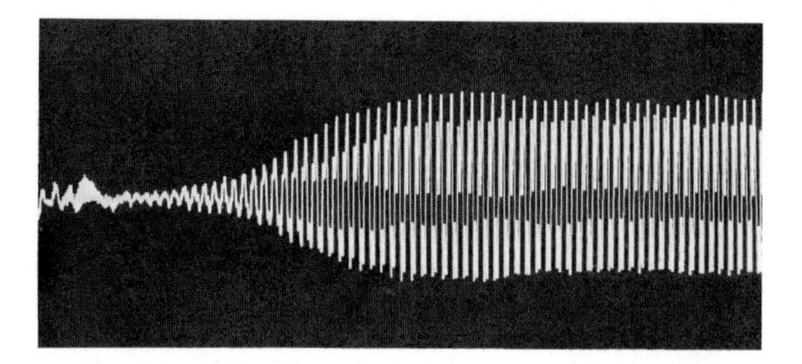

2.20 *The start of a note on a metal flute; the total time of the part shown is about 1/10 second*

In general the sound produced by these so-called 'edge tones' is smooth and has the quality which can only readily be described as 'flute-like' as opposed to the much harsher tone produced by reed instruments. We shall return to discuss this point in rather more detail later after discussing the reeds, but it may help even now to remark that the smoothness really stems from the fact that the eddies themselves move smoothly from one side to the other. One can imagine the jet of air 'waving' back and forth from the inside to the outside of the tube rather than switching over suddenly.

Although relatively smooth, the sound produced by a flute or whistle takes time to develop. First the air has to build up to the right velocity and there is often a characteristic 'breathy' sound just before the note proper starts. Then there is an argument of the kind discussed for the tuning fork; the first eddy travels up the inside of the tube as a compression. When it reaches the end it suddenly finds itself free to expand and – as always with mechanical systems – the expansion goes too far, air from the tube moves out, and hence an expansion wave travels back along the tube until it reaches the edge. If it happens to reach the edge just as the next eddy is ready to travel up the inside again the oscillation in the pipe will build up. If not the sound produced will not be musical and the player has to adjust the velocity until the eddies do match. Even with a skilled player this may take a few cycles and this adds characteristic features to the build-up period of the note. 2.20 shows the oscillograph trace of the start of a note on a metal flute which is fairly typical of this family.

(e) Blowing a reed – for example, bagpipe and bugle

The wide range of varieties and types of reeds used in musical instruments provide a field day for experts in classification, but we shall not attempt to discuss anything other than the broad principles. It is always interesting (though sometimes misleading!) to speculate on the origins of words and of instruments. Does the name 'reed' arise from the fact that the earliest examples were made from hollow reeds? Without pursuing this interesting speculation at all we can nevertheless find a very useful starting point for a discussion of musical reeds by considering two simple types that can be made from hollow reed stems, corn stalks or, perhaps more readily in these days, from paper drinking straws. To make the first, use either a stalk cut just above a knot so that the top of the pipe is closed or bend over the end of a paper straw (2.21 a). Then slightly bend the straw with a razor blade and make a rectangular cut about 2 or 3 mm wide and about 2 cm long. To blow this reed the whole top section of the pipe must be placed in the

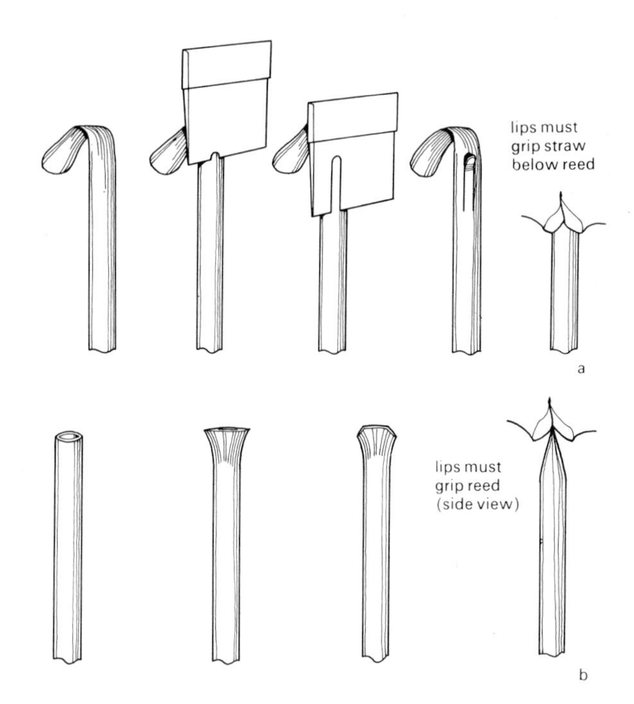

2.21 a and b *Simple reeds from drinking straws*

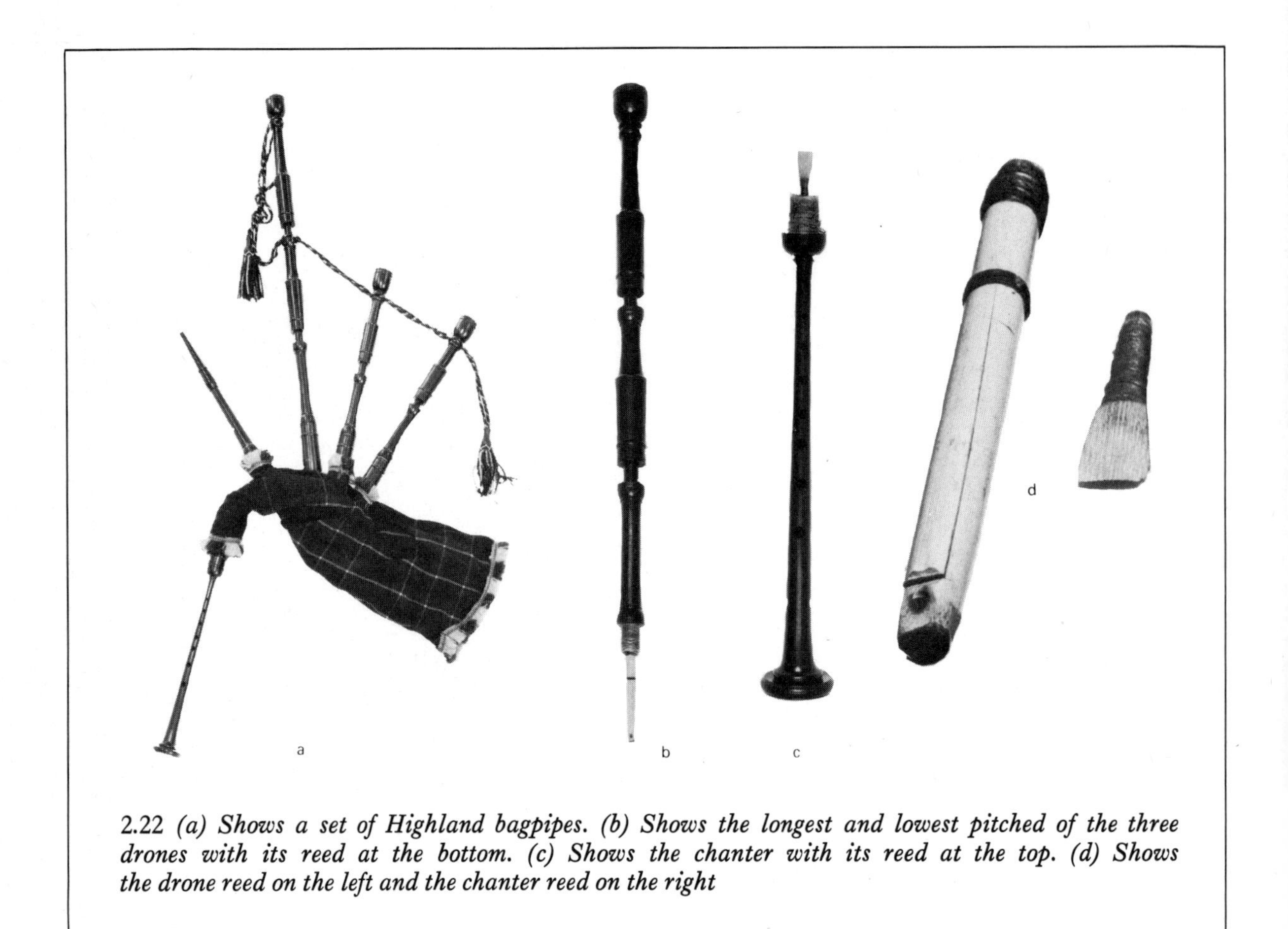

2.22 *(a) Shows a set of Highland bagpipes. (b) Shows the longest and lowest pitched of the three drones with its reed at the bottom. (c) Shows the chanter with its reed at the top. (d) Shows the drone reed on the left and the chanter reed on the right*

mouth with the lips touching the stem below the base of the reed. If it does not produce a note immediately it can usually be coaxed to do so by very slightly lifting the reed and bending it up so that when at rest the hole in the side of the pipe is not quite closed. The second type of reed is even easier to make (2.21 b), merely flatten the end of the straw and cut each corner at about 45°, removing only about a millimetre or two at each side of the top. This reed is pressed between the lips and blown fairly hard.

The first of these is surprisingly similar to the drone reeds of the Highland bagpipes and the second to that of the chanter on which the melody is played (2.22) (though it is not placed between the lips but inside a hollow tube). In the bagpipes they are, of course, much more substantial and made of hollow bamboo, but the principles in both cases are identical.

Though rather more sophisticated in construction, clarinet and saxophone reeds operate in a manner which is not far removed from that of the first type described and the oboe resembles the second. The first is a single reed and the second is double; both are what are technically described as heterophonic rather than idiophonic. All this means is that they are capable of vibrating at a wide range of different frequencies and can match the air column to which they are attached. The lips of the player of any brass instrument would be described as a double heterophonic reed. Idiophonic reeds are usually of springy metal – such as in the mouth organ, harmonium or piano accordion – and have one specific frequency defined principally by the shape, size and material of the reed.

Whatever type of reed is under consideration one can think of it as a device for turning on and off the flow of air through it. For some reeds the flow may actually stop completely for a certain fraction of the cycle, as for example in the chanter of the bagpipes; in others the flow of air is merely varied in volume but never quite stopped, as for example in the mouth organ – in which the tongue of the reed is very slightly smaller than the hole in which it fits so that the hole can never be completely closed.

Idiophonic reeds are sometimes used without any amplification or modification, though the skilled harmonica player uses the cavity formed between his hollow palms as an amplifier and modifier of the basic tone and this will also affect the beginnings of the notes he produces. Heterophonic reeds, however, depend very much on the pipe or vessel to which they are attached for determining their pitch. The operation is somewhat similar to that described in the last section except that the interaction or argument between the reed and the pipe tends to be much more violent. The slow-speed tape-recorder has already been mentioned as a means of revealing audibly the irregular start of a bagpipe practice chanter. 2.23 shows the corresponding oscilloscope trace. At first the reed opens and closes irregularly and

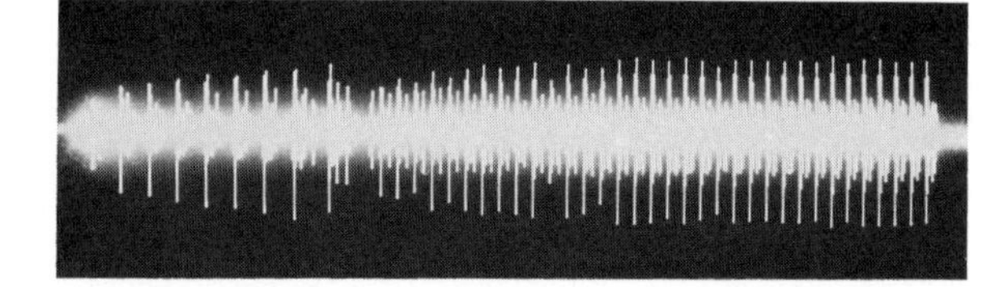

2.23 *The trace of a note of a bagpipe practice chanter starting up from the left. The whole trace lasts 1/5 second*

puffs of air travel along the tube and are reflected back. The time taken to travel up and back is fixed by the tube length and the returning pulses react on the reed and gradually force it to open and close regularly in time with the returned pulse as can be seen on the right of the trace. This whole trace lasts about a fifth of a second, so we can see that it takes about a tenth of a second for the reed to settle down.

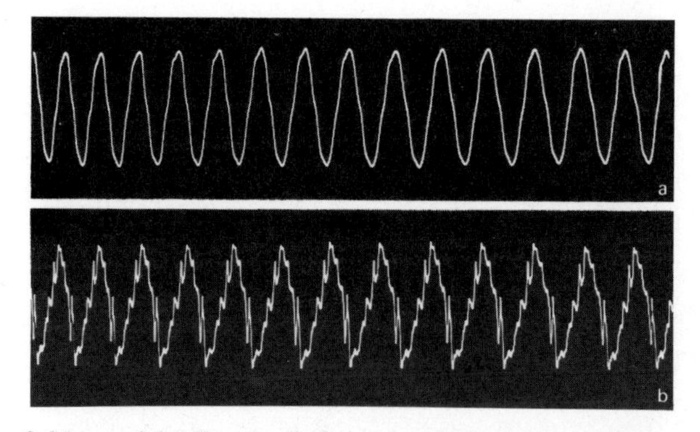

2.24a and b *Trace of the note produced by a 15-cm length of tubing: (a) with a recorder mouthpiece at one end and closed at the other; (b) with a clarinet mouthpiece at one end and open at the other*

Why do reeds sound so much harsher than instruments using edge tones? Compare 2.24 a and b. These are oscilloscope traces of the tone produced by the same 15-cm length of 1-cm diameter bakelite tube first with a recorder mouthpiece setting it into vibration using edge tones and secondly with a clarinet mouthpiece. Notice the great increase in higher-frequency components when excited by the reed. Again this point will be elaborated in the next section. (In order to obtain notes of the same pitch, when the recorder mouthpiece was used the bottom end of the pipe was closed; with the clarinet reed it was left open as the reed itself behaves as a closed end. In both, therefore, there was effectively one closed and one open end.)

One important group of reed instruments that has only been mentioned in passing has the capability of producing both smooth 'rounded' notes as well as harsh strident ones. The brass family make use of the lips of the player as a double heterophonic reed, but because the lips can be varied in shape and tension by the marvellous muscular system which controls them, the range of qualities, both deriving from the different starting transients and the modes present in the steady state, is extremely wide. A bugle sounding 'Reveille' has a totally different quality from that of the haunting long notes of 'Last Post'. In appendix IV similar differences are exemplified in the extract from Mozart's Third Horn Concerto (K447).

ANOTHER LOOK AT HARMONICS

We cannot delay any longer taking a more scientific look at the effects of the method of initiating sounds on the mixture of modes or harmonics generated.

First, a word on terminology. The terms 'mode', 'harmonic', 'overtone' and 'partial' are all used, to some extent, interchangeably in the literature. The meanings which we shall adopt in this book are as follows.

'Mode' refers to any specific pattern of vibration associated with a specific frequency and may be used when only one pattern is exhibited or when mixtures occur.

'Partial' is very similar to mode but is usually used when mixtures occur and refers rather more to the frequency or pitch. For example, one might refer to the 'upper partials' of a bell, meaning the frequencies of the higher and more complex modes of the bell which form part of the total sound.

'Overtone' is similar to mode but applies only to patterns higher than the main or fundamental one and is usually used when a single mode is predominant. For example, a tin whistle blown hard to produce a note an octave above the normal one might be said to be producing its first overtone.

'Harmonic' is the only term that has a completely specific scientific definition. It can only mean an integral multiple of some fundamental frequency and the number of the harmonic is always that of the integer involved.

We can perhaps summarise the meanings by referring to the note produced by overblowing a clarinet. This, for reasons hinted at towards the end of chapter one, has a frequency three times that of the fundamental note – that is, an octave plus a fifth higher. The fundamental mode is the first partial or the first harmonic, but it is not an overtone at all. The note produced by overblowing would be the first overtone, but it would be the second mode or second partial and it would be the third harmonic; in this case the second harmonic should not exist if the simplified view described so far were completely valid; since we are only concerned with terminology here we will not pursue the realities any further until later.

In order to try to understand why the method of initiation has an effect on the modes present it will be helpful to introduce another experiment using a pendulum. 2.25 shows an experiment which is quite easy to carry out or to visualise.

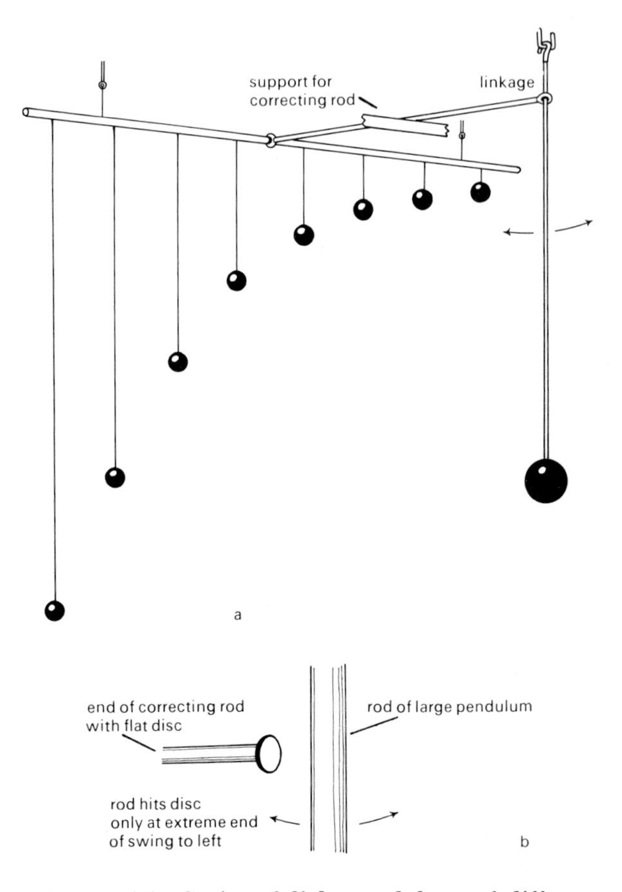

2.25 a and b *Series of light pendulums of different lengths driven by one heavy one: (a) with a constant drive, (b) with a 'push' once every forward swing*

Seven or eight different pendulums with wooden bobs, say about 3 cm in diameter, each with a different length of string are suspended from a single bar which itself hangs from two short pieces of string so that the whole bar can swing. The lengths of the strings are chosen so that, say, four of the pendulums have harmonically related times of oscillation and the other four are unrelated. For example, one might choose frequencies

of A = 0.45 Hz E = 1.5 Hz,
B = 0.5 Hz F = 1.85 Hz

C = 0.65 Hz G = 2.0 Hz
D = 1.0 Hz H = 2.05 Hz

Then A, C, F and H are unrelated (except, if we want to be really precise, that they are the ninth, thirteenth and forty-first harmonics of a vibration of frequency 0.05 Hz!), whereas D, E and G are the second, third and fourth harmonics of B. The approximate lengths of the pendulums would turn out to be

A = 1.22 m E = 0.11 m
B = 0.99 m F = 0.072 m

C = 0.58 m G = 0.062 m
D = 0.25 m H = 0.058 m

A further pendulum with a very massive metal bob and a steel rod hangs near to the rest and is adjusted to have a frequency corresponding to that of B. A rigid metal rod links a point fairly high on this pendulum with the bar on which all the others are hung. When the heavy pendulum is set swinging it will carry on for a long time because it is so massive and will force the bar carrying the other pendulums to follow its motion exactly. In practice all the pendulums will move a little bit in a rather erratic sort of way, but B will build up to a very high amplitude indeed. This is, of course, the familiar phenomenon of resonance. Why do the other pendula fail to build up? Let us consider D as a typical example. The main pendulum starts to swing in one direction which we will call forward and will give D a push also in the forward direction. The natural frequency of D is twice that of the heavy driven pendulum, however, and hence by the time the heavy pendulum is in the middle of its backward swing D will have swung back and be moving forward again. The 'push' will then be at the wrong movement in the cycle and will tend to slow D down. If one follows this kind of analysis with each pendulum we find that for all except B the driver spends as much time pushing in the wrong way (out of phase) as it does in the right way (in phase) and hence their swings never build up.

Suppose now we disconnect the rod joining the driver pendulum to the rest and arrange it so that it receives a short push just at the moment when the main pendulum reaches the extreme of its swing towards the rest but at no other time (2.25 b). What happens now? B will again build up because it receives pushes at regular intervals. D will also build up now because, although it will make twice as many swings as B, it will receive a push at the right point of *every other* swing and will not be pulled in the wrong direction at all. E will receive a push of the right type every third swing and G every fourth. Thus the four harmonically related pendulums will now swing vigorously. The other four will still not swing very much because the pushes they receive will not stay in step with their natural periods and, though they may move from time to time, there will be no resonance.

Finally let us ask ourselves what would happen if we gave the bar supporting the pendulum one big push and then left it free. Clearly, this time, all eight would start to swing.

What is the significance of all this for our present discussion? First we see that if the driver or primary vibrator is itself vibrating smoothly rather like a pendulum (performing what a physicist would call simple harmonic motion) it is only likely to excite a single mode or partial having exactly its own frequency. Thus the edge tones of whistles and flutes in which the air jet waves gently back and forth over the wedge tend to produce a relatively pure tone with only one main mode being excited. (We used the recorder (1.13) as a method of producing pure tones.)

However, when we move to reeds in which the air is cut off for quite a large part of the cycle and we have a series of short puffs of air the resemblance is much closer to that of the second example above in which the driver pendulum gave a regular series of short pushes. The result is that all the various harmonic modes are also excited – leading to the familiar 'reedy' tone. Notice, however, that non-harmonic partials are unlikely to be excited if the reed is operating at a steady frequency.

What does the single push on the bar correspond to? Clearly the striking of a bell or drum, and it is not surprising, therefore, that the result is to excite both harmonic and anharmonic partials.

Again we have picked out extremes. In order to find out what happens with bowed strings we should need to replace the driver pendulum by some rather odd mechanical device that moved the bar much more rapidly one way than the other. Without going into details the model will perhaps suffice to indicate that it is at least reasonable to expect a different selection of partials from that produced either by smooth or pulse excitation.

Our pendulum model is being used as a frequency analyser and, if we had a very large number of pendulums indeed, we could use it to tell us what modes are likely to be excited by any given basic signal. Our real time frequency analyser does just this. Often instead of saying that these are the 'frequencies or modes likely to be excited by a given signal' we say that the given signal 'can be analysed into these components' or indeed is 'made up of these components'. The remarkable mathematical fact then emerges that, if one takes a set of waves of the frequencies and amplitudes of the components of which a given note is found to be made up by using the analyser, they do in fact add up to give a wave corresponding to that note. This principle – called Fourier Analysis and Synthesis – is of great importance throughout physics and it is hoped that the models and discussions given here will at least give the reader an indication of what it can do and of how its results can be used.

ANALYSIS AND SYNTHESIS

Some interesting philosophical questions arise. Is the wave really made up of these components? Which comes first – the wave analysed into components or components added up to form a wave? The answer – as often happens in physics – is that these are just alternative ways of looking at the same thing. Our oscilloscope tells us how the wave varies with time; the analyser tells how the wave varies with frequency. To round off the discussion we give in 2.26 four different wave forms produced electronically together with the corresponding frequency analysis. The first is for a single pure tone, the second is for a regular sequence of narrow pulses of the same basic frequency – rather like that produced by a reed – the third is for a so-called saw-tooth wave form rather like the motion produced by bowing, and

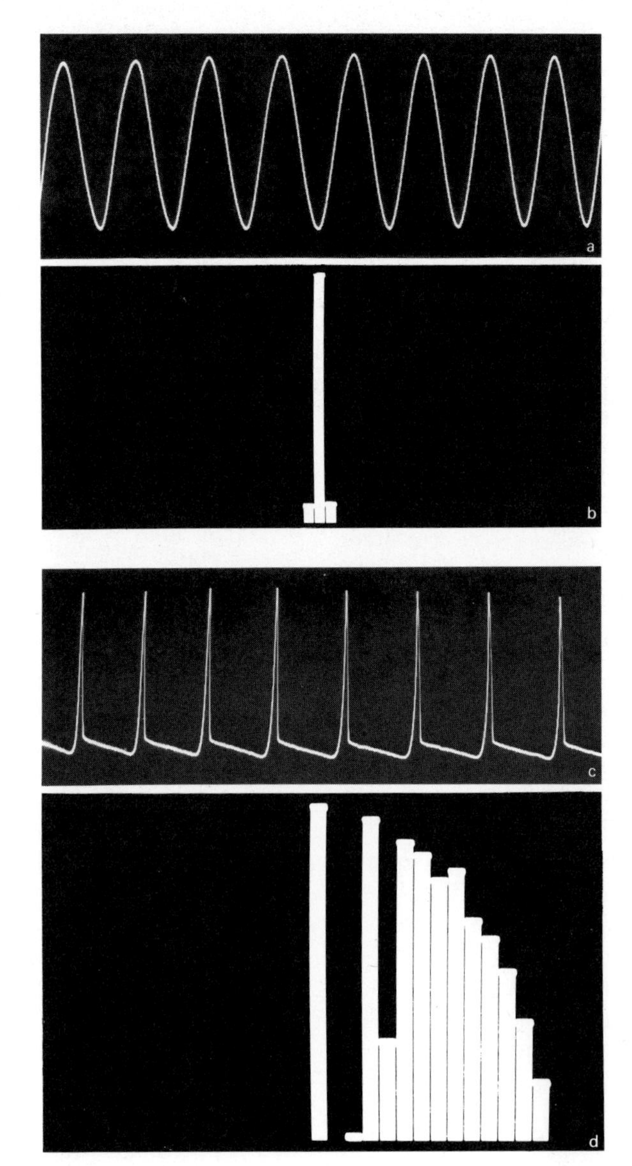

2.26 a–d *Traces and analyses of various waves: (a) pure tone; (b) analysis of (a); (c) sequence of narrow pulses of the same frequency as (a); (d) analysis of (c)*

the final one is for a single pulse like the hammer striking a bell. The analyses correspond closely to those predicted by our earlier pendulum experiment.

Finally we will introduce a mechanical device which helps us to see what happens when we add waves together. The principle is illustrated

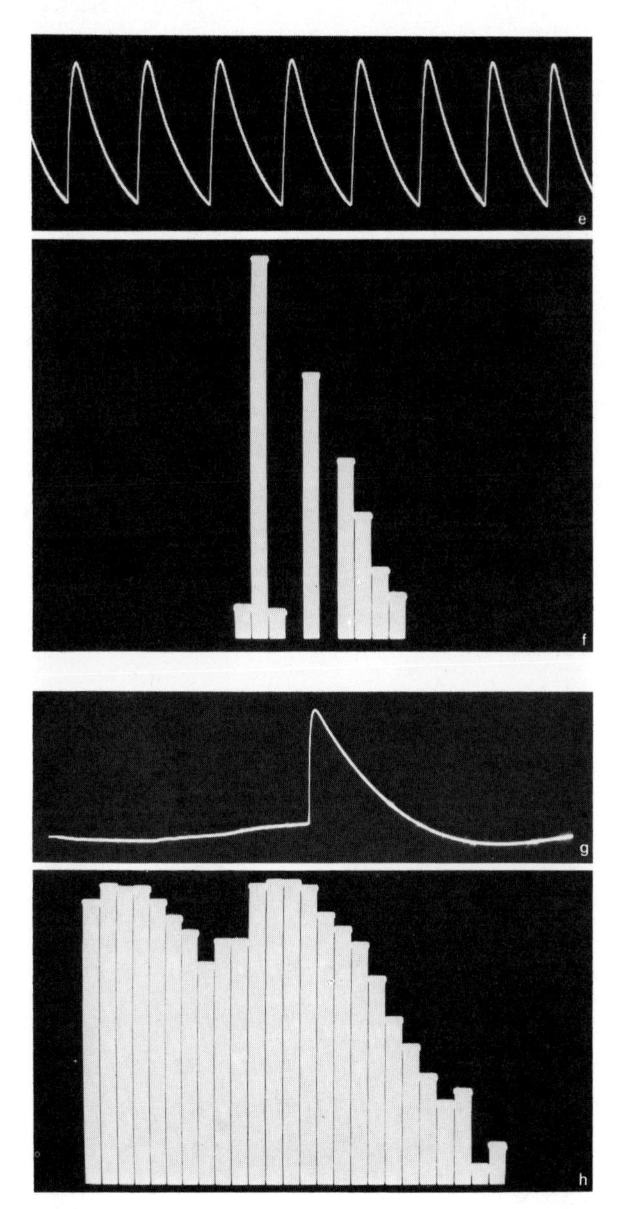

2.26 e–h *(e) 'saw tooth' wave of same frequency as (a); (f) analysis of (e); (g) single, rather broad pulse; (h) analysis of (g)*

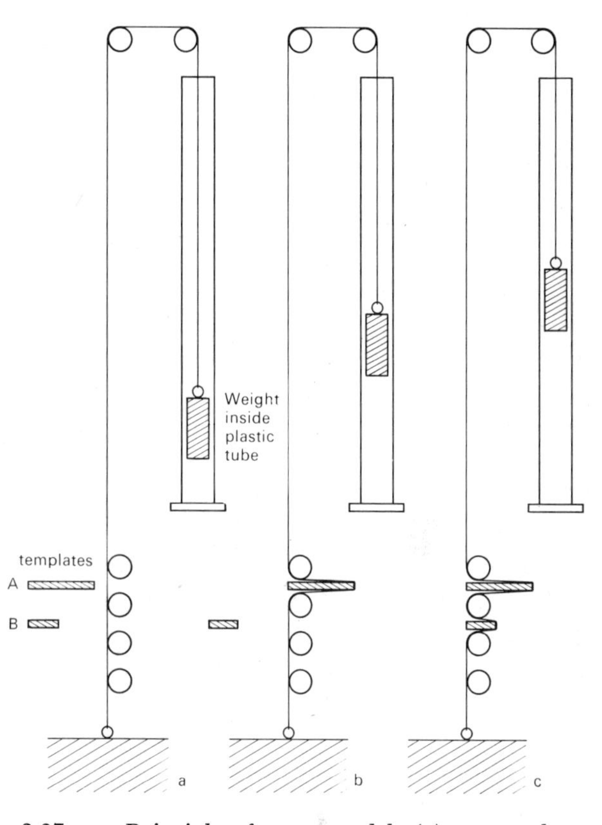

2.27 a–c *Principle of wave model: (a) no templates in position; (b) one template inserted; (c) two templates inserted*

in 2.27; a weight hangs on a string which passes over pulleys in such a way that if a sheet of plastic is inserted at A the weight rises by an amount which is clearly related to the amount by which the plastic is pushed in. If a second piece of plastic is inserted at B the final position of the weight will depend on the sum of the two displacements due to A and B separately. The model shown in 2.28 has fifty of the weights and up to ten plastic templates can be inserted, each representing a different wave form. In 2.29 a a single template, representing a wave of a single frequency, has been inserted. In 2.29 b a second has been added corresponding to a wave of twice the frequency and we can see the result of the addition. The templates are shown in 2.29 c, and in 2.29 d is shown the result of placing the second template a little to one side of its initial position. This movement represents a change in the phase

2.28 *The complete wave model*

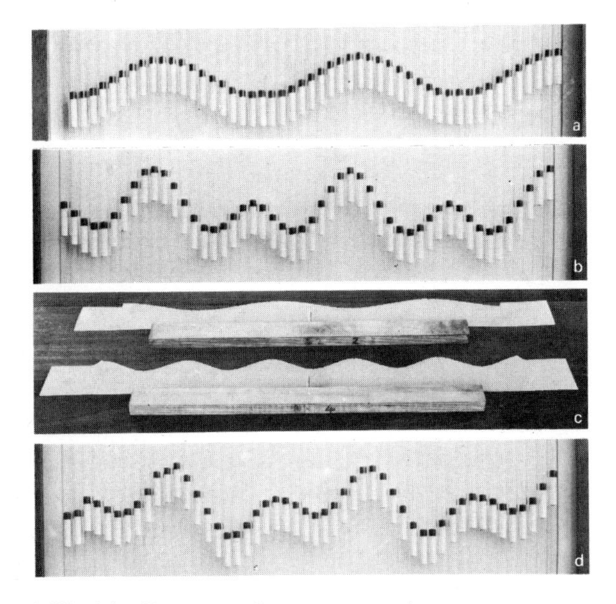

2.29 *(a) One template representing a pure tone inserted. (b) Template representing a tone an octave higher (2nd harmonic) added. (c) The templates used for (b). (d) As for (b) with the templates for the upper note moved to represent a phase change*

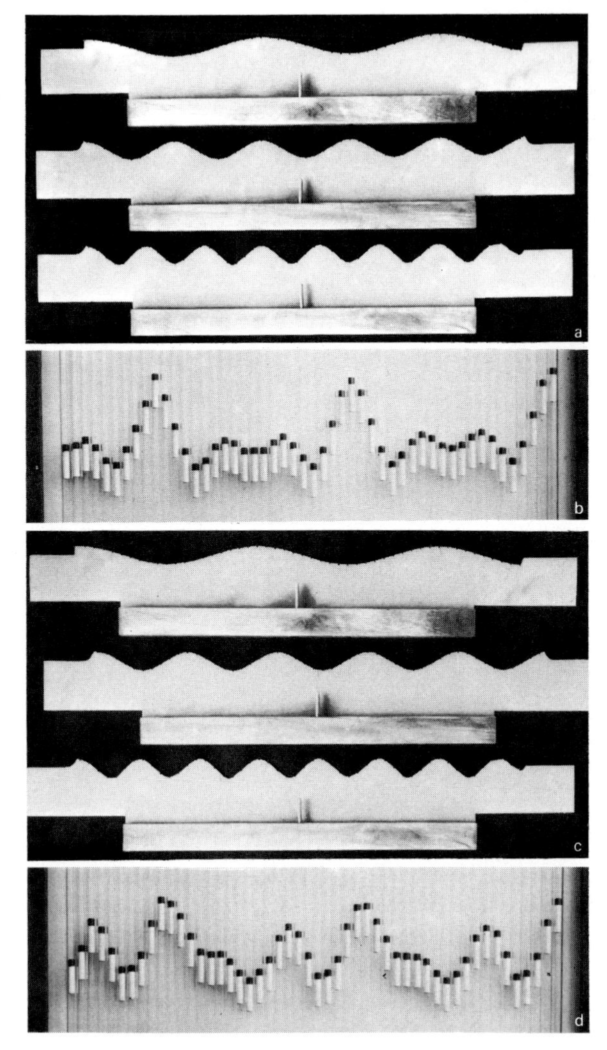

2.30 *(a) Template representing waves with frequency ratios 1 : 2 : 3. (b) Shape produced when all three are inserted. (c) The middle template has been moved to the right to represent a phase difference. (d) The result when the templates are inserted as at (c)*

(the relative time at which the waves pass a particular point) and is clearly significant in determining the wave shape. It turns out, however, that for these simple harmonic additions the ear cannot distinguish any difference, for example, between the sound corresponding to 2.29 b and 2.29 d. In chapter six we shall see that there are circumstances in which the ear is able to recognise phase differences – but that complication must wait. In 2.30 additions of three components with frequency ratios 1:2:3 are shown, again with the templates and showing the effect of changing the phase of the middle component.

WHAT ARE THE MAIN POINTS OF THIS CHAPTER ?

Though it has taken a considerable time and a wide range of illustrations, there are only three really vital points that emerge from this chapter that must be clearly understood before we may proceed. The first is that the very beginning of each note of music or each syllable of speech is of great importance in helping us to recognise it quickly – probably more important than the rest. A good illustration of this is that changes in the vowels (the steady parts) of words have far less effect on meaning than changes in consonants. Whether we say trumpit, trampet, trompit or trumpet, the words are recognisably the same; but crumpet or drum kit are instantly recognised as something totally different!

The second point is that the starting transient depends both on the mechanism used to initiate the vibrations and also on the interaction between the various components of the system; a classical example of the first would be the difference between plucking and bowing a violin, and of the second the comparison between the musical attack at the beginning of a note on the saxophone and the same note on a clarinet; the reed system is virtually identical but the body shape totally different.

The third point is that the interaction between the exciter and the main vibrator can actually change the quality of the steady part of the note as well as just the beginning. The example to note here would be the enormous difference in tone between a tin whistle and a bagpipe practice chanter; the length and general shape of the main vibrator – the pipe – is not very much different, but the edge-tone excitation of the whistle gives a totally different quality from that of the chanter.

Some clear examples of oscillograph traces of the transients of real instruments playing normally can be seen in the two parts of appendix IV. For example the piano chords o, p, q of bar 10 of the Kegelstatt Trio show examples of transient behaviour which are particularly striking in the white-on-black enlargements of the traces. In the Horn Concerto there are several instances of horn transients, for example in notes o, p, q and r of bar 156, where the notes are staccato and of increasing loudness.

Now, having summarised the fascinating consequences of initiating a note, we are ready to consider the next stage in its development as it grows and changes.

Charles Taylor demonstrating the apparatus shown in 3.27 during his Royal Institution lectures

GROWING AND CHANGING

IT'S NO USE IF YOU CAN'T HEAR IT

The most useless musical instrument I have ever made is a one-string fiddle which produces so little sound that you can hardly hear it at all! 3.1 shows how it is made; the bridge is a steel rod which is mounted in rubber and consequently the vibrations of the string are not transmitted to the wood. The result is rather like that for a tuning fork which is held in the hand; so little air is in contact with the vibrating string that only very small disturbances are produced. The one-string fiddle can be made useful, however,

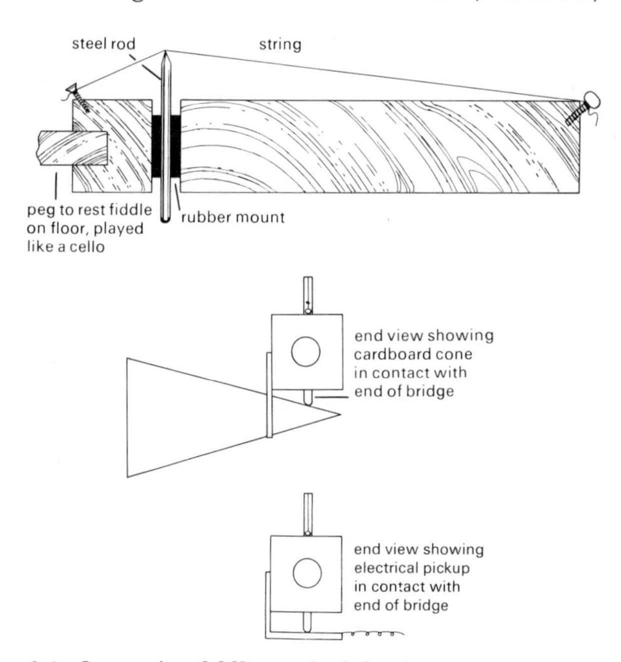

3.1 *One-string fiddle – principle of construction*

in two ways. We can add an acoustical or an electrical amplifier. The acoustic amplifier consists of a cardboard cone fixed so that the free end of the 'bridge' touches it near its apex; cones of different size give not only different degrees of loudness but also different qualities of sound. Electrical amplification is easier to control, however, and the first step is to convert the mechanical vibrations into electrical ones with a pick-up. The kind of pick-up used on an electric guitar, a small microphone or a very cheap gramophone pick-up cartridge will do quite well and is placed in contact with the bridge. The output from it is then fed to an amplifier and loudspeaker. Immediately the room can be filled with sound, and by varying the tone controls on the amplifier we can change the quality of the sound produced. This amplification, which is so necessary in order to make the instrument audible, brings with it the complications of starting up that we mentioned in the last chapter (even the electrical amplifier and the loudspeaker system have a response time to each note, though it is probably relatively short), but it also brings the complication that it is almost impossible to make a sound louder without changing its quality; a great deal of the skill and craftsmanship of instrument makers is devoted to the design of the mechanical amplifiers in order to give all the desirable tone qualities and to eliminate the unpleasant ones. The scientific principles underlying the design form the main subject-matter of this chapter.

The way in which the amplification varies with frequency is sometimes called the formant characteristic. In some instruments it stays fixed whatever note is being played and in others it may be different for every note. Once more the human voice falls into a class apart since it exhibits both kinds of formants, some of which are concerned with the recognisable quality of a particular person's voice and others which are concerned with the vowel sounds which have the same formants whoever says or sings them at whatever pitch. Formants thus will play an important part in this chapter. Just as the sound takes time to get started it also takes time to grow and its loudness may vary considerably before it finally fades away. It is common to talk about the overall shape – as displayed on a slow-speed oscilloscope – as the envelope; it is the outline within which the waves are packed. We shall devote a little time to this aspect of growing and decaying towards the end of the chapter.

A much more scientific way of changing the tone of the one-stringed fiddle is to use an electronic gadget called a 'spectrum shaper'. First we should explain the use of the word spectrum. In the study of light we refer to the analysis of a particular beam of light into its component frequencies as spectrum analysis. The eye interprets variations in frequency as changes in colour and what we know as the rainbow is really a spectrum analysis of the sun's light. The idea is carried over into acoustics and we even talk about 'white noise' when we mean sound which is capable of exciting responses at all possible frequencies in the audible range, just as white light contains components at all frequencies in the visible spectrum. Noise which contains a greater proportion of lower frequencies than of higher is called 'pink' noise, again by analogy since the red part of the visible spectrum is the lower frequency end. (The visible spectrum ranges from about 400 million million Hz at the red end to 750 million million Hz in the violet – i.e. almost one octave in acoustic terms.) The frequency analyser introduced in the last chapter is thus sometimes called a spectrum analyser; a spectrum shaper is any amplifier which permits one to vary the spectrum. Any amplifier with bass and treble controls is a spectrum shaper of a kind – but the type referred to above is much more specific. 3.2 a shows what it looks like – the slides increase or decrease the amount of amplification at thirty different frequencies and the shape made by the positions of the slides is crudely the spectrum produced. 3.2 b shows the spectrum, as displayed on the real time frequency analyser, produced by a setting similar to that shown in 3.2 a.

This is done by feeding into the spectrum shaper noise which would, without the shaper,

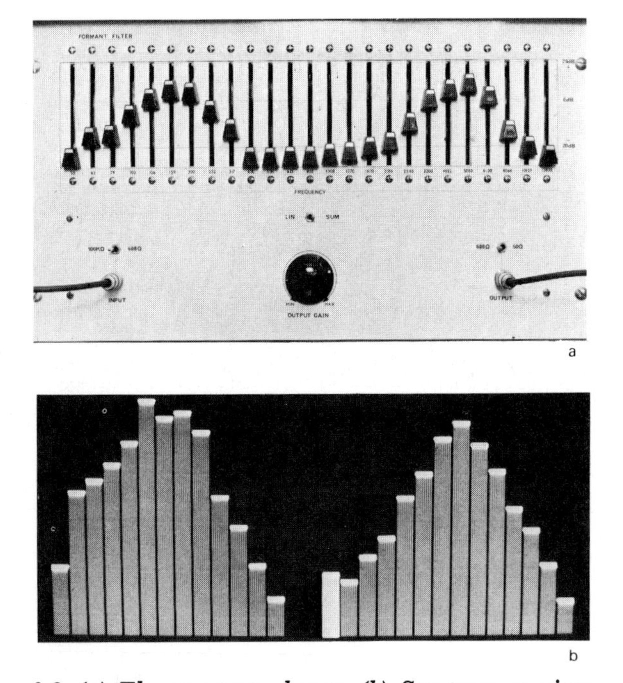

3.2 *(a) The spectrum shaper. (b) Spectrum as given by the setting shown in (a)*

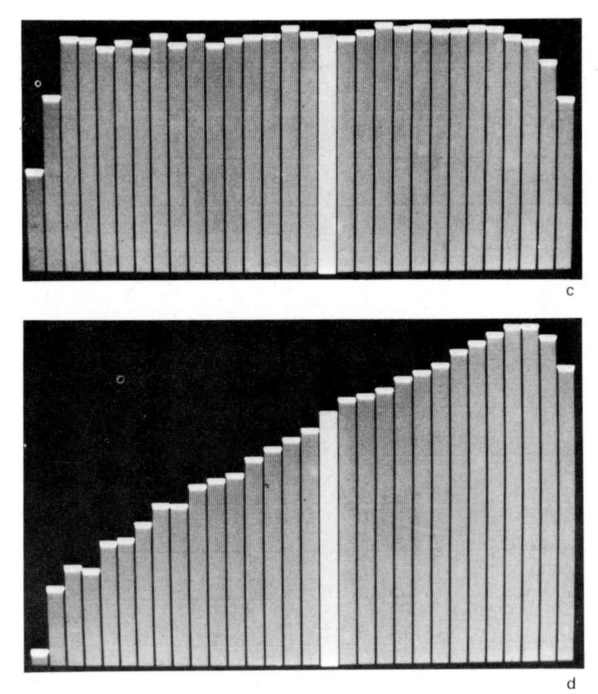

3.2 *(c) 'Pink' noise. (d) 'White' noise*

produce uniform bars on the analyser (3.2 c) – i.e. having equal amounts of all frequencies. One can then see exactly what the shaper does to the spectrum. (A purely technical point must be put in here for electronic experts; because the analyser sorts the spectrum into bands which are all one-third octave in width rather than of

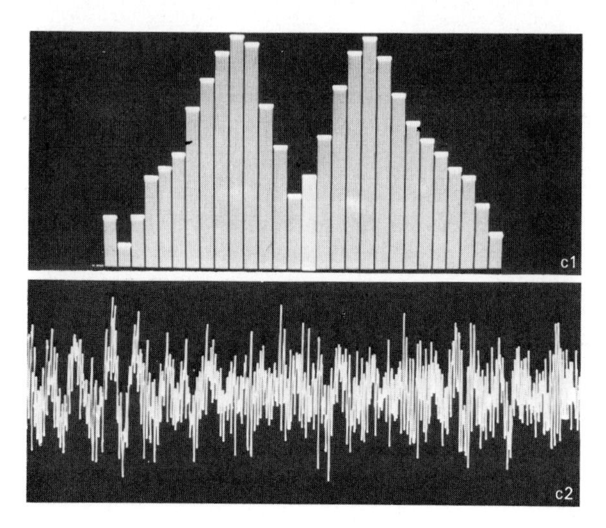

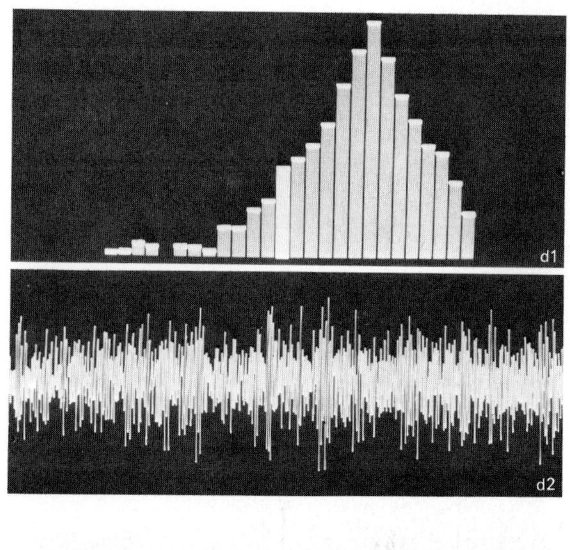

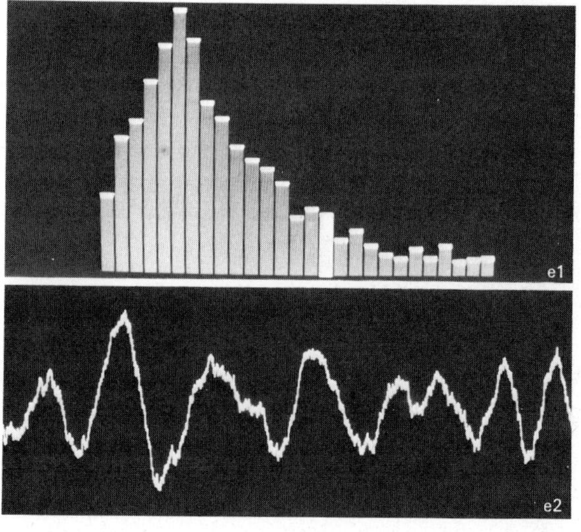

3.3 *(a)–(e_1) Analyses of pink noise into five settings of shaper. (a_2)–(e_2) Wave forms of 3.3 a_1–e_1. N.B. In the analyses of 3.2, 3.3, 3.4, 3.5 and 3.6 the 1 kHz vertical strip is brightened for reference purposes*

equal *frequency* width this means that as one goes up in frequency so the frequency width of each band increases. If white noise were used, the increase in width means that the energy per one-third octave band increases as they move to higher frequencies on the analyser (3.2 d). To compensate for this it is necessary to feed 'pink noise' into the analyser to obtain the flat response and hence pink noise is used in all experiments displaying the response spectrum of devices.)

The combination of spectrum shaper to set the rough pattern required and frequency analyser to measure what has actually been achieved provides a very good basis for starting our study of amplification and formant characteristics. It will give us a useful background if we perform a few purely scientific, or electronic, experiments first of all. In 3.3, 3.4, 3.5 and 3.6, five different settings of the spectrum shaper a, b, c, d and e are used. In 3.3 the signal fed in is pink noise and so 3.3 a_1, b_1, c_1, d_1 and e_1 are the actual spectra imposed by the shaper. In the remaining figures, a_1–e_1 show the frequency analysis of the wave emerging from the system and a_2–e_2 show the shape of the wave form emerging. Since the first setting is level, then 3.3 a_2, 3.4 a_2, 3.5 a_2 and 3.6 a_2 represent the wave forms actually fed in. Table 3.I also attempts to describe the sounds and the changes that occur in words. The effects are very marked.

So much for our electronic demonstrations. But what are the mechanical devices that are actually used in musical instruments or in the human vocal system that may be the source of amplification and formant characteristics? They fall into three main categories and we shall consider each separately. First we shall consider resonance – the steady build-up of oscillations by applying impulses at one of the natural fre-

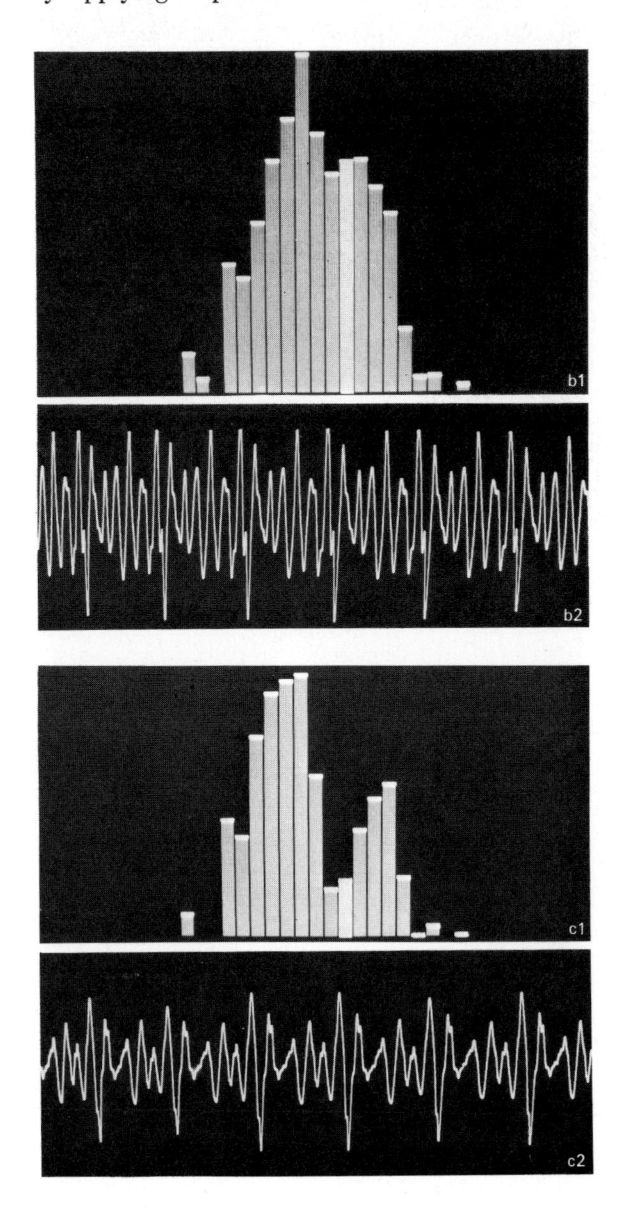

a1

a2

quencies of the system – in long thin things, mainly pipes and tubes, but with a passing reference to strings. Secondly we shall consider resonance in much more complicated shapes built up of two-dimensional plates and three-dimensional boxes. Thirdly we shall consider the ways in which sounds are radiated from instruments such as through the flared horns of the brass instruments or out of the side holes in the woodwind. The formants of the voice are so important that they are given a separate section, although really their origin places them in our second category.

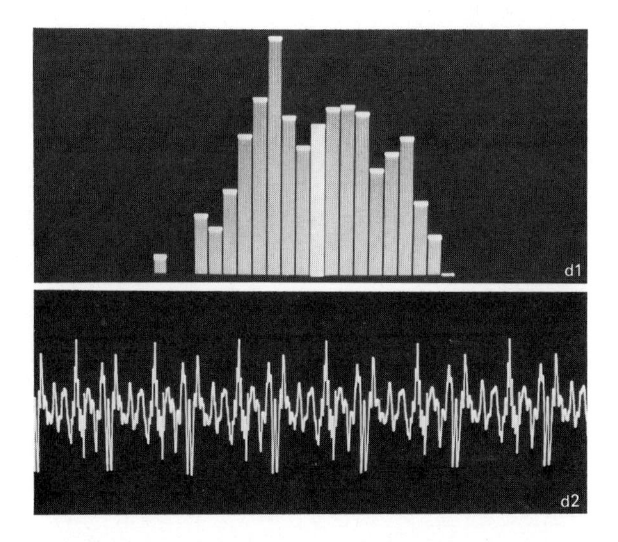

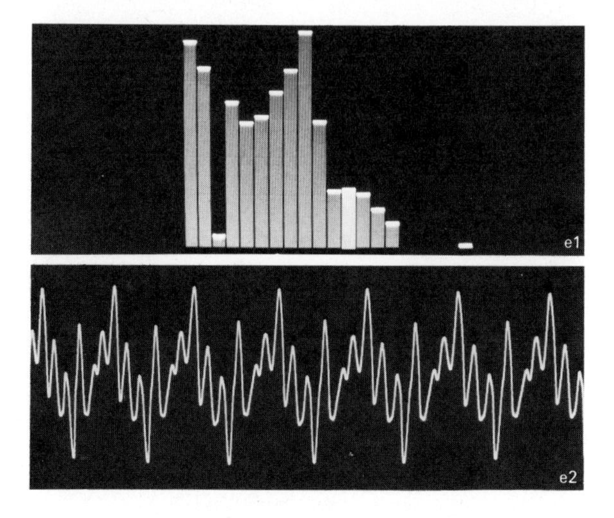

3.4 *(a_1)–(e_1) Analyses of bassoon note into five settings of shaper. (a_2)–(e_2) Wave forms of 3.4 a_1–e_1*

RESONANCE IN PIPES AND TUBES

Towards the end of chapter one we began to look at modes of vibration of long thin vibrators and in chapter two we introduced the idea of resonance and saw that a system which has a natural frequency of vibration will pick it out and respond to it when motion involving a complex mixture of frequencies is imposed on it. We now need to bring these two ideas together because in at least two of the great families of orchestral instruments – the woodwinds and the brass, not to mention organ pipes – long thin columns of air form the vibrating system which responds to the initiation or excitation of a reed or edge tone. There are many different possible approaches, many of which involve complicated mathematics, but we shall try to analyse the problem in a very simple way. It could be argued that it is over-simplified, and certainly it does not account for some of the finer points of the results observed. It is my contention, however, that it does not present any *incorrect* ideas; where there are defects they arise because the presentation is incomplete.

First we must think about what happens when a wave is reflected in some way. The interesting point is that what happens depends on the type of boundary at which the reflection occurs. A fixed boundary sends the pressure wave back exactly as it came, whereas a free boundary changes the compression to an expansion or rarefaction. This statement certainly needs clarification. Perhaps one of our mental pictures will help. Imagine a queue of children waiting to go in to a party or something of that sort; the door is closed, but they are all eager not to miss a moment so there are no gaps in the queue. A new arrival runs up and collides with the end of the queue, so squashing those already waiting. The 'squash' will be passed along until it reaches the child standing next to the very solid doors and he clearly cannot pass the squash on any further; his reaction is to try to recover by pushing back the last but one person in the queue

TABLE 3.I

SETTING OF SHAPER	*Level*	*Single hump peaked at 1 KHz*	*Double hump 300 Hz + 2500 Hz*	*High peak 4 KHz*	*Low peak 100 Hz*
SOUND FED IN					
Pink noise	Like hiss of steam	Also like steam but not quite so high-pitched	Like a very loud whispered 'Ay' as in Hay	Like ultra-high-pressure steam escaping	Low roar like distant traffic
Bassoon at about 88 Hz	Like a natural bassoon	Like a bassoon played less expertly – slightly rough	Like someone singing 'Ay' through a comb and tissue paper	Very thin and harsh but still recognisably the same note	Mellow hum
Saw tooth at about 200 Hz	Humming sound like a machine	Same pitch but more 'reedy'	Like someone singing 'Ay' through a comb and tissue paper	Thin buzz like a bluebottle fly	Mellow hum
Male voice 'Ah'	Natural sound	Little change	Like saying 'Ah' or 'Ay' through comb and tissue paper	'Ah' very thin and far away as on a bad telephone line	Sounds like someone saying 'mmm'

and so the 'squash' is passed back to the beginning again. In other words, at a rigid boundary (the door) a squash is reflected as a squash.

Now consider what would have happened if – just at the crucial moment – the doors had been opened, i.e. we had a free boundary. When the 'squash' arrived at the child next to the door there would be nothing to stop him moving forward and he would probably be propelled quite vigorously into the room. Now suppose that all the children were holding hands; the end child would move into the room until he began to pull the next one in and so on. It is not now a 'squash' that is reflected but its opposite, a 'pull', extension or rarefaction. Thus we find the second part of our rule; at a free boundary the pressure in the wave is reversed.

Now we should apply this principle to the air in a pipe. Suppose we begin with a pipe open at both ends, that is it has free boundaries at both ends. Now imagine that we send a single brief compression pulse into one end; it will travel to the other end, expand into the free air and more air will follow it from the pipe; it will be reflected as a 'rarefaction' which will travel back to the first end again. When it reaches its starting point again, air from outside will tend to fill up the rarefaction and rush into the tube, hence a compression pulse will travel back and the cycle will be repeated. Now let us imagine that we supply a second pulse just at the precise moment that our first pulse begins its second journey. Clearly we have the repeat conditions for resonance – as with the pendulum or the child's swing already mentioned – and the vibration can build up. But notice that it only works satisfactorily for short pulses – just as, in our pendulum experiment in the last chapter (2.25), the harmonically related pendulums were set swinging only when the drive worked for a brief part of the cycle. The frequency of the note will depend on the time taken for the wave to travel twice the length of the tube – out and back again – and so, for example, in a pipe of length 0.6 metre the wave would travel a total of 1.2 metres, between each new compression. At a velocity of 300 metres per second (0.3 metre per ms) this would clearly take 4 ms and so there would be 250 compressions per second sent down the tube, or a frequency of 250 Hz would result. It could

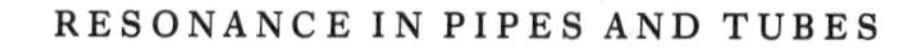

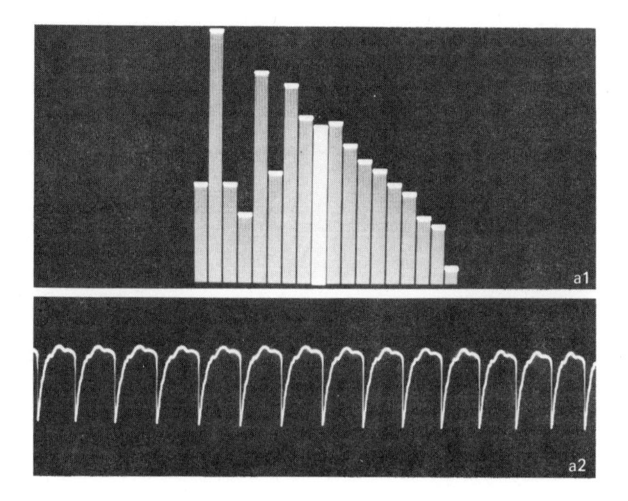

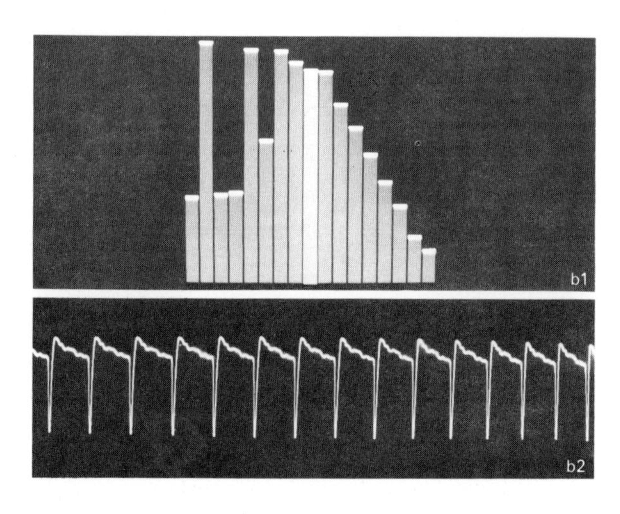

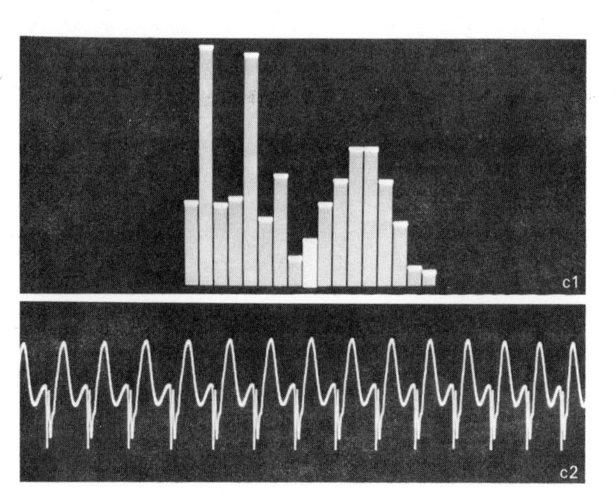

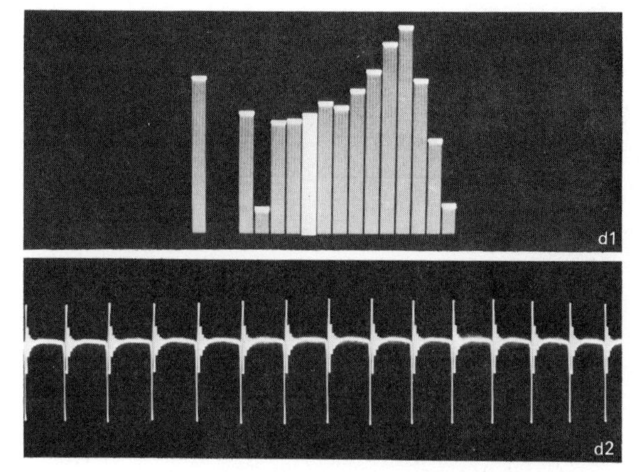

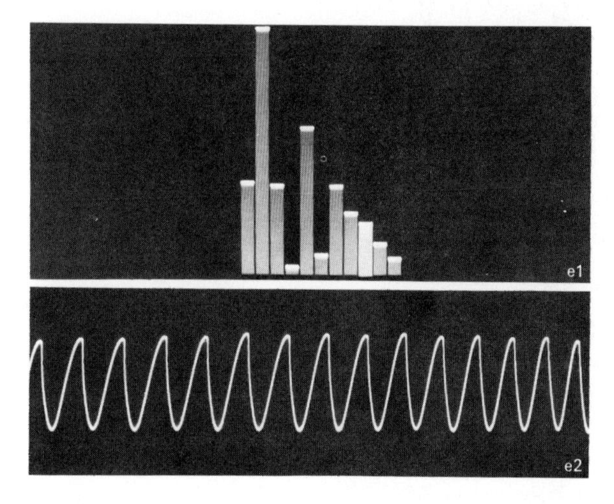

3.5 *(a_1)–(e_1) Analyses of saw-tooth wave into five settings of shaper. (a_2)–(e_2) Wave forms of 3.5 a_1–e_1*

also build up just as well if we applied the pulse at twice the frequency, or three times, or four times, and so on. Only every second, every third, every fourth, etc., pulse would be used to reinforce the first pulse, but the intermediate ones would also reinforce each other and so we would find just as effective a build-up. For example, for the fourth mode the first pulse would complete its outward and return journey in time to be reinforced by the fifth and again by the ninth; similarly the second, sixth, tenth; the third, seventh, eleventh and the fourth, eighth, twelfth would reinforce and so on. There would now be one pulse every millisecond and the frequency would be 1000 Hz, which is the fourth harmonic of 250 Hz. 3.7 may help to make this clear.

The very important point to notice is that the pulses 'take no notice of each other'. Waves just pass through each other without creating any permanent change. (Scientifically speaking this is known as the principle of superposition.) A moment's thought makes it plain that this *must* be so or we could not communicate. I can shout to my friend across the field while a cow moos

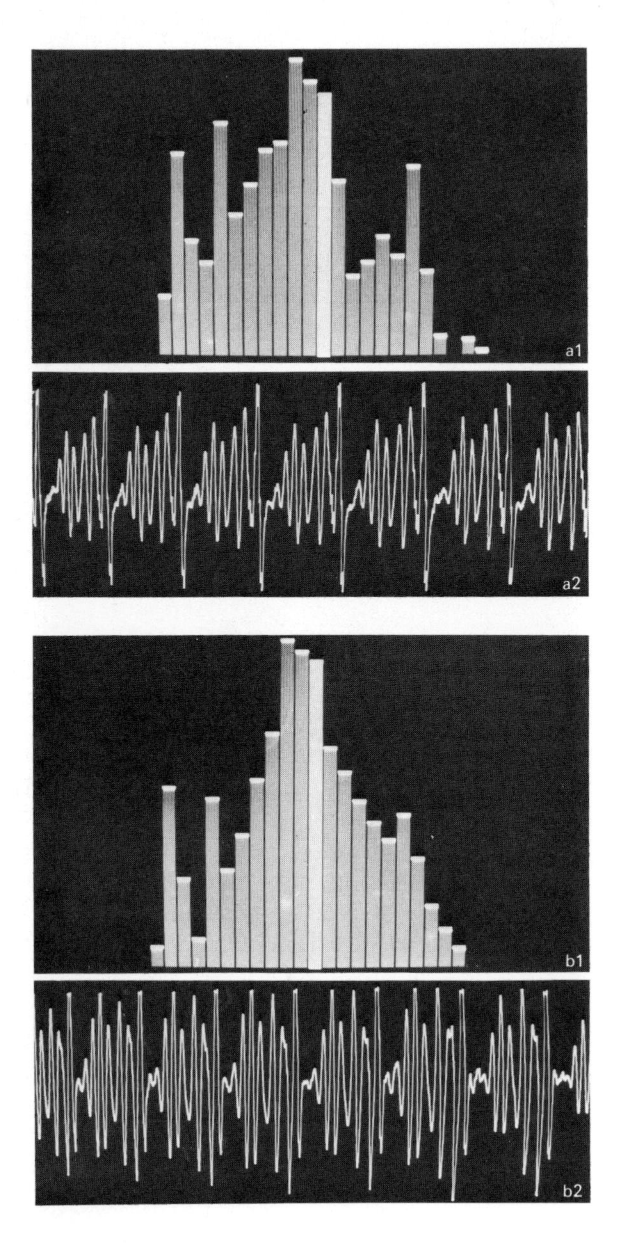

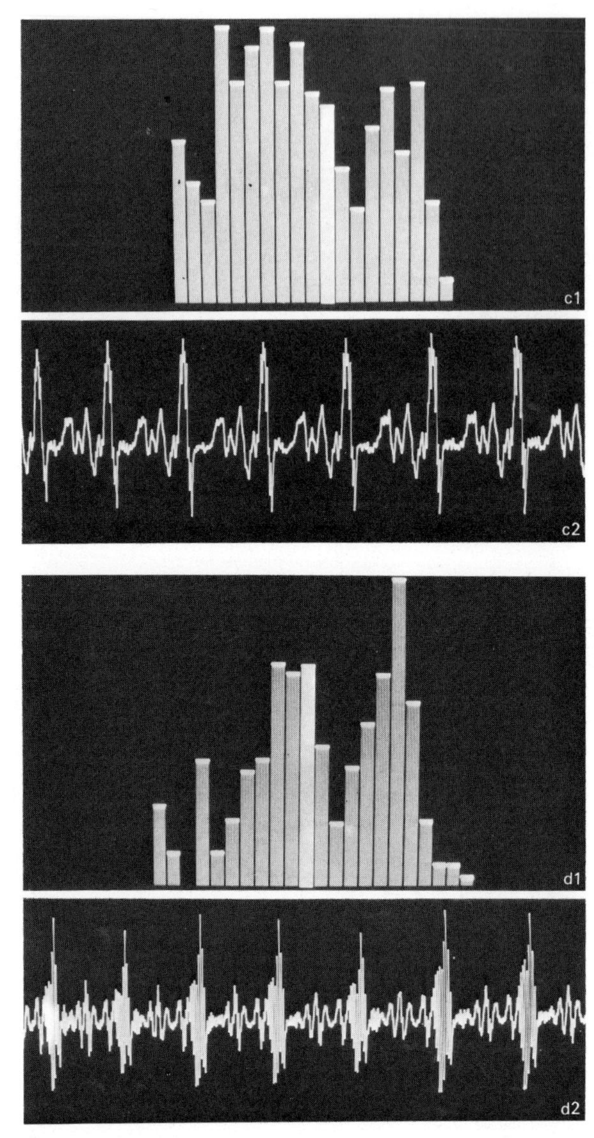

to its calf and a skylark calls to its mate, and though we can all hear each other we can still disentangle the separate sounds – they do not become a composite shout – moo – tweet! Nevertheless, going back to the point made right at the beginning of the book, the air pressure can only have one value at one point and so, if we make measurements at a particular point, it will be the sum of the waves that we measure. Thus in 3.7 a if we were to measure the pressure at the right-hand end of the tube after 0.002 s

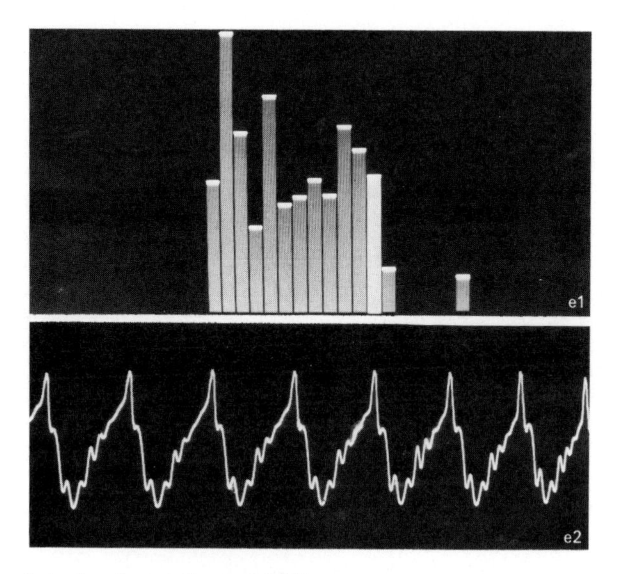

3.6 *Analyses of vowel 'Ah'*
(a_1)–(e_1) with five settings of shaper
(a_2)–(e_2) Wave form of 3.6 a_1–e_1

we should find that the incoming compression adds to the returning rarefaction to make the pressure equal to that of the outside air. This is really, of course, the scientific reason for the reversal of pressure on reflection. If the net pressure difference were not zero we should either find more and more air crowding into the pipe or more and more leaving it, whereas of course in practice the average amount remains the same.

The idea of the principle of superposition – that waves pass through each other without inflicting any permanent change and yet momentarily one measures the sum of their two effects – is a difficult one to grasp but is extremely important. A picture that may be useful is that of a man walking up an escalator which is going down. If the speeds are equal and opposite, the net result to an observer who can only see his upper half would be the same as though the climber were stationary on a fixed staircase. Also by varying the speed of the escalator the outside observer could be made to think that the man was going at any speed either up or down, although the man himself still kept climbing at the same steady pace and, if the escalator were stopped, he would be seen to be walking upwards undisturbed. Probably the most striking example of superposition observed every day is that of light waves. In a room full of people light waves are scattered in all directions and each person can see each other person or object in the room simultaneously without any one set of waves upsetting any other set!

Now, having spent some time convincing you of the importance of the principle, I am now going to warn you that we have recently come to believe that part of the secret of the tone of both wind and string instruments is tied up with the fact that within some of the mechanical components – such as the reeds of wind instruments or the bow-string interaction in fiddles – the principle is not obeyed; they behave in what physicists would call a non-linear way. This really means that there is not a simple, direct relationship between what you do to the system and the way it responds – but more of that later.

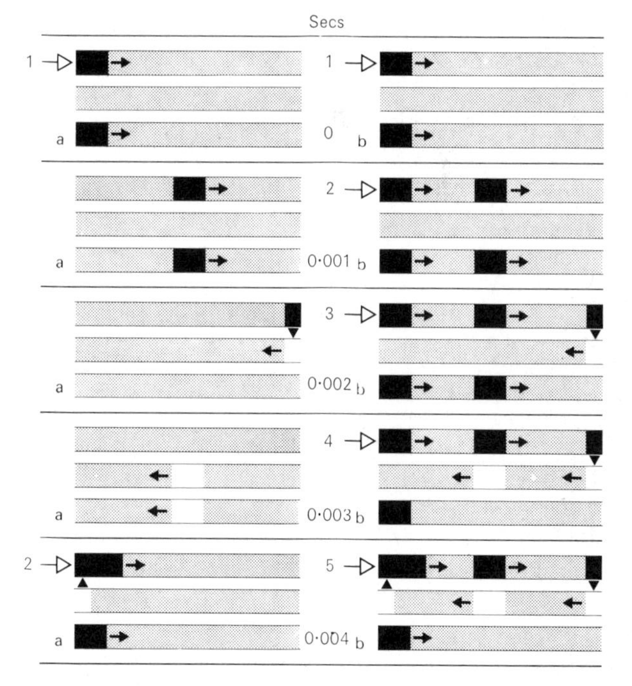

3.7 *On the left (a) the pipe is fed with short pulses at its fundamental frequency and on the right (b) at four times this frequency. In each example the top diagram represents pulses travelling to the right, the second the reflected pulses travelling to the left and the bottom diagram is the resultant*

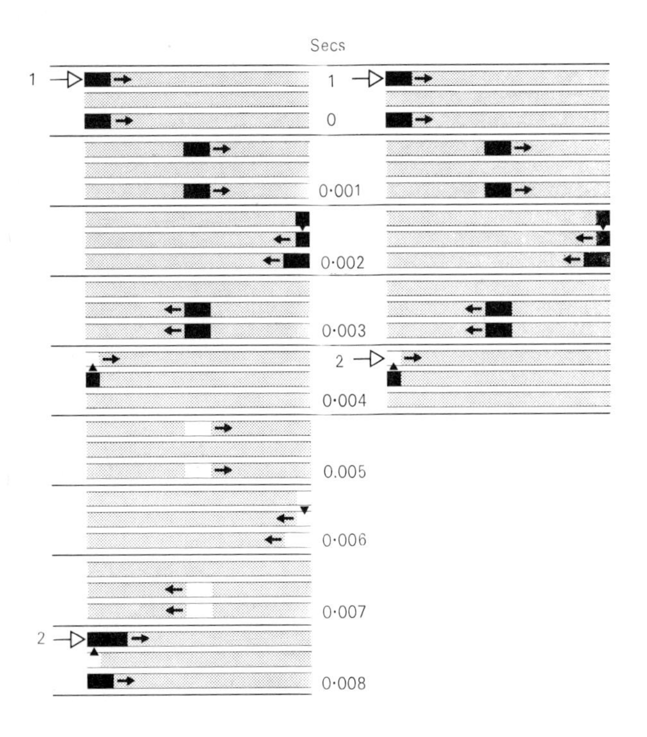

3.8 *As 3.7 with pipe closed at the right-hand end*

Let us turn to pipes closed at one end and open at the other. If we send a compression pulse down from the open end it will be reflected as a compression. When it returns to the beginning, however, it meets an open end and so starts its second cycle as a rarefaction, is reflected from the closed end as a rarefaction and finally after *four* trips along the tube (out, return, out, return) is reflected at the open end as a compression. This time we need to send a pulse down at intervals corresponding to *four* times the time taken to travel the length of the tube. In other words the fundamental mode has *half* the frequency or is an octave lower than for the open-ended tube of the same length. You can easily check this principle for yourself if you take a short length of tube and blow across the end – the note may be a bit 'breathy' but should have a recognisable pitch. Now put your thumb over the bottom end and the note drops an octave. If you cannot produce a note this way take just the top section of a treble or descant recorder – that is, the part with no finger holes in it. Blow the recorder and then place the palm of your hand over the open end to make a fixed boundary. In 3.8 a the length of the tube is still 0.6 metre, but the wave now must travel 2.4 metres between each new compression and this takes 8 ms, giving a frequency of 125 Hz.

Now we come to a more complicated piece of reasoning. What happens if we send pulses at twice the rate of the fundamental into the open end of a pipe which is closed at the other end? The second compression pulse will be setting off just as the first pulse is starting the second half of its journey as a *rarefaction*. And this will happen with all the pulses and so the net result is that they do not build up. Three times the rate is all right, however, as the first pulse will start its journey as a rarefaction half-way between the second and third compression pulses. If you follow this reasoning carefully you can see that modes with frequencies 1, 3, 5, 7, etc., are the

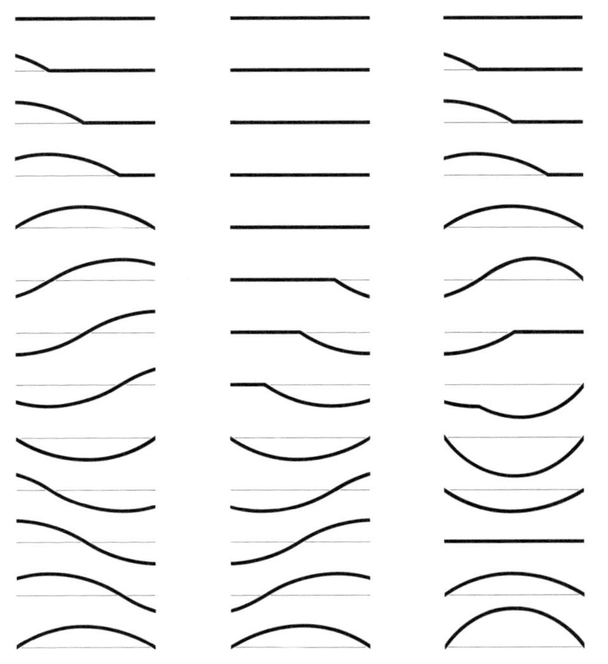

3.9 *Graphs of the pressure changes at 1/8 cycle apart as a wave builds up in a tube open at both ends. The left-hand column is the wave moving from left to right; the middle column is the reflected wave moving back; and the right-hand column is the graph of the net result.*

only possible ones. 3.8 b may help. For both these examples we started by considering a single pulse travelling along and considered the conditions that would lead to any particular mode becoming established as a steady note, but we only looked at the patterns at certain time intervals. Before leaving this point we should perhaps consider what actually happens while a particular mode is being built up. 3.9 illustrates the case for the fundamental mode only of a pipe open at both ends. In order to set up this mode *alone* we need to apply a pressure at the end which varies 'sinusoidally', that is in the way that the pressure at a single point varies in the wave corresponding to a completely pure tone. (We introduced this idea in the last chapter in talking about the pendulum model.) Its frequency must be such that one complete cycle is completed in the time it

3.10 a *Pressure-change graphs for a pipe closed at the right-hand end with a sine wave of a frequency corresponding to the fundamental if the pipe had been open. The first four columns show the traverses of the tube, two in each direction, the final column the net result. In the end everything cancels out*

3.10 b *As for 3.10 a but fed with a sine wave of* half *the frequency, showing that resonance now builds up*

takes the wave to travel from one end of the tube to the other and back again. In the figure a graph of the pressure is drawn at intervals of one-eighth of a cycle. The fascinating point is that we have two equivalent pictures of what is happening: we can either consider the two sets of waves travelling back and forth as in the first two columns with the compression becoming a rarefaction and the rarefaction becoming a compression each time the boundary is reached; alternatively we can consider the sum of the two in the third column which shows that the total effect viewed from outside is as though the pressure at each end stays constant and equal to that in the room while the pressure in the middle changes from high to low. Now notice that the graphs in column 3 of the pressure in the pipe open at both ends exactly resemble the motion of the rubber rope in 1.25 a and so we see yet again how closely interconnected all these points are.

The diagrams for the pipe closed at one end are more difficult to draw and we need five columns because you will recall that, for the fundamental mode, the wave has to travel the length of the tube four times. In 3.10 a we have attempted to show what happens if a tube closed at the far end is fed with a pure sine wave of the frequency appropriate to the fundamental mode of a tube *open* at both ends. You will see that, although disturbances of various kinds occur at first, by the time the wave has travelled the full four lengths everything cancels out and no resonance results. In 3.10 b the frequency of the sine wave is halved and a vibration mode with a pressure minimum at the open end and a point of maximum variation of pressure at the closed end builds up as predicted.

The resultant wave forms in column 3 of 3.9 or column 5 of 3.10 b are sometimes called stationary or standing waves, since *to an outside observer* the wave does not seem to be moving along in either direction. But it must be stressed that we are discussing two alternative ways of visualising what is happening rather than two opposing principles.

The existence of the points of maximum and minimum pressure variations in a tube can clearly be made visible by means of a little device invented by Mr Gibbs of Bristol University. It consists of a tiny microphone of the type used for deaf aids mounted at the end of a long rod. A transistor amplifier unit housed, with its batteries, at one end of the rod amplifies the signals picked up by the microphone until they are powerful enough to light a small pea lamp which is mounted near the microphone. The result – often referred to as a 'magic wand' – is a system which will respond visibly to pressure variations; a sensitivity control is also incorporated in the handle. The circuit, modified slightly from that given by Mr Gibbs, is shown in 3.11. 3.12 a shows a glass tube effectively open at each end and excited by a small loudspeaker at one end which is fed with pure tones of various frequencies. By sliding the wand in and out of the tube the points of maximum and minimum pressure variation can be found. In 3.12 b and c time exposures were taken and the rod slowly moved along the tube. 3.12 b shows the third mode and 3.12 c the fifth.

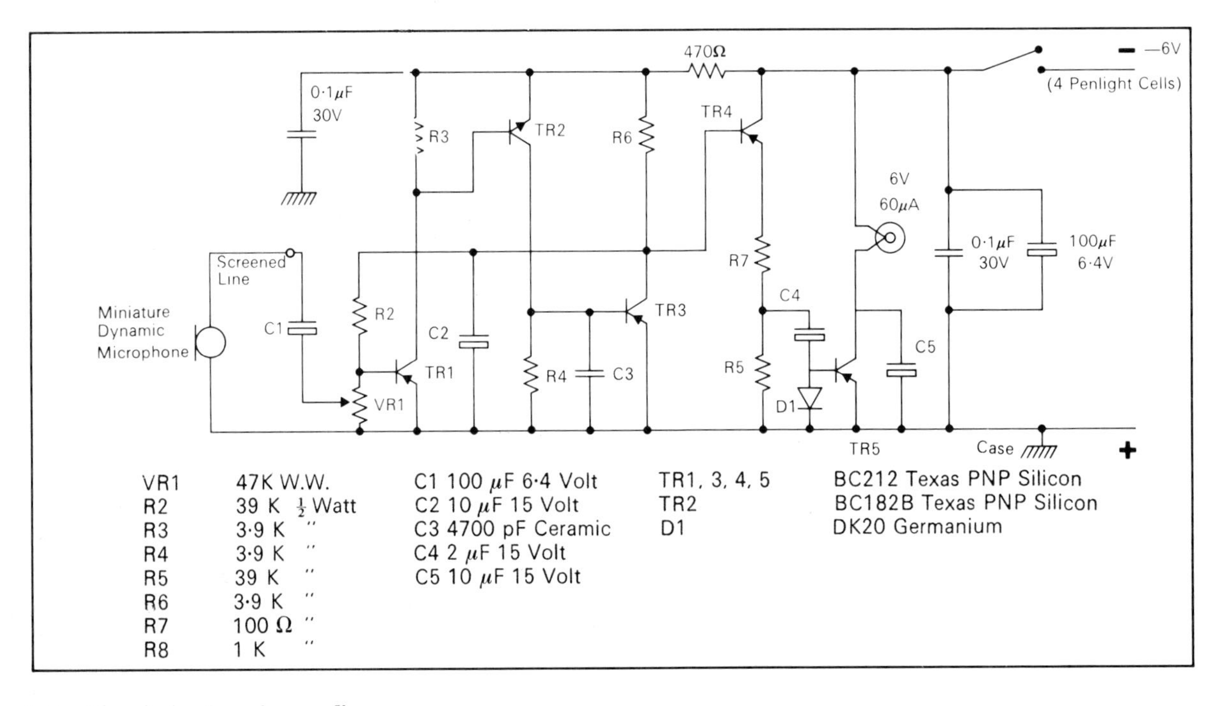

3.11 *Circuit for 'magic wand'*

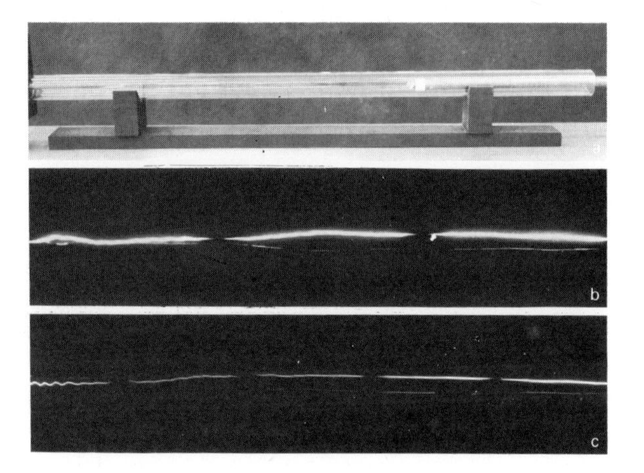

3.12 *(a) Magic wand inside tube excited by small loudspeaker at the left-hand end. (b) Time exposure as wand is moved along tube with the third harmonic excited. (c) Time exposure as wand is moved along tube with the fifth harmonic excited*

Now we should pick out two vital points from this discussion that are of great significance, but one of which, oddly enough, was not stressed at all in textbooks on sound and music until very recently. The first and well-recognised one is that, as with the pendulums of 2.25 in the last chapter, a long thin vibrator such as a pipe or string has harmonic modes of vibration and can resonate in response to a pure tone corresponding to one of its modes; it can respond to a complex of tones, picking out any frequencies present in the complex which correspond to modes, of the pipe; or it may even respond to a sequence of sharp pulses, as from a reed, at the fundamental or any harmonic frequence and these will excite all the higher harmonic modes. The second point often overlooked is what happens if we try to feed the pipe with strong pulses at a *slower* rate than that of the fundamental.

Let us suppose that we take a pipe open at both ends for which a note of 120 Hz will excite its fundamental note: that is, it takes 1/120 second for a wave to travel from one end to the other and back again. Suppose now that we feed the pipe with *sharp* pulses at the rate of 60 Hz. Think of the child's swing: we don't necessarily have to push *every* swing in order to build up a large angle of swing, a push every other one will do – but the swing still swings at its natural frequency and not at the rate at which the push is administered. We can see therefore that the 60 Hz puffs will excite the fundamental note of the pipe at 120 Hz! This is sometimes called excitation by a sub-harmonic frequency, and once the steady state sets in one can imagine that, though the main note will be 120 Hz, every other puff will be slightly bigger because it has been reinforced and our ears may recognise a 60-Hz component emerging from the end as well. But it is very important to remember that we are talking about excitation by sharp pulses – not by pure tones.

Now for the complication that is so unexpected and tended to be neglected until a few years ago. Suppose we apply sharp puffs at 180 Hz. This is *not* one of the natural frequencies of the pipe as we have so far understood the term: but it would correspond to giving every alternate compression of the 360-Hz mode – the third harmonic of the tube – a push. Hence not only can we excite the third harmonic by puffing at 180 Hz but, by the same argument as before, every alternate compression will be a bit bigger and we have a component of 180 Hz coming out as well! This is the 'one-and-halfth' harmonic of the pipe and makes nonsense of some of the things we said earlier! Scientists are used to this kind of paradox, however, and we can take it in our stride. Remember that it only happens when we are exciting the tube with sharp pulses and clearly the resonance will not be so strong, nor the note produced so loud, as the normal mode. The whole sequence of frequencies at which a pipe can be excited even weakly are called the privileged frequencies of the pipe. *Table 3.II* shows those for the pipe just discussed compared with its true harmonic modes. One can extend the idea to exciting every third compression – though now of course the result will be much less clearly defined and even every fourth may work. Notice in the table how the true harmonic modes (printed in bold type) crop up among *all* the sets of privileged frequencies – which we have called first order, second order, third order according to whether every compression, every second, every third, etc., are rein-

TABLE 3.II

True harmonics	**120**	**240**	**360**	**480**	**600**	**720**	**840**	**960**	**1080**	**1200**
First-order privileged frequencies	60	**120**	180	**240**	300	**360**	420	**480**	540	**600**
Second-order privileged frequencies	40	80	**120**	160	200	**240**	280	320	**360**	400
Third-order privileged frequencies	30	60	90	**120**	150	180	210	**240**	270	300
Fourth-order privileged frequencies	24	48	72	96	**120**	144	168	192	216	**240**

Privileged frequencies for a pipe open at both ends, approximately 1.25 metres long. Only the first ten harmonics are used and true harmonics are printed in bold type

forced. These privileged frequencies explain some of the interesting features of reed instruments and they have a great influence in the brass family. This introduction is probably over-simplified and a proper understanding of it makes us realise that, as already mentioned, reeds of all kinds behave in a non-linear way and cause departures in the system from the principle of superposition. Experimentally, however, the effects are easy to demonstrate. Take a straight length of electrical conduit or similar tubing 1 or 2 cm in diameter and about $1\frac{1}{4}$ metres long. Fit it with a tuba mouthpiece – or if not available a connector of the type used to fix a hosepipe on to a tap will sometimes do. Try to blow bugle notes and you will find that there are many frequencies at which your lips are pulled into vibration – though some are much stronger than others. The lips act roughly as a closed end and so on traditional theories one would normally expect only odd harmonics; the sequence you might now theoretically expect is

60	75	84	100	105	108	132	135
140	150	156	165	**180**	195	204	210
220	225	228	255	260	270	285	**300**

etc.

You are unlikely to obtain many of the lower ones and only those in bold type – the true harmonics – will be strong. *Table 3.III* shows the origin of these various frequencies.

Another quite startling experiment – which can only be done successfully with a tuba mouthpiece – uses two pieces of hose: one, say, 2 metres long and the other, 0.68 metre long. The fundamental of the long tube at about 37 Hz can be excited fairly easily with the lips very relaxed. But now change the mouthpiece to the shorter tube and try to make the same lip vibrations; you should find that the same low note can be produced. But how can a pipe only 0.68 metre long give a note of wavelength 8 metres? The answer is that it is a privileged frequency. The fundamental of the short tube is about 110 Hz and one-third of this – the privileged frequency for which every third pulse is reinforced – is about 37 Hz!

Professor Bouasse seems to have been the first to recognise the existence of privileged frequencies, but – probably because he wrote in a challenging forthright style of French and made his views on the shortcomings of his scientific colleagues all too clear – his work was not recog-

TABLE 3.III

True harmonics	**60**	**180**	**300**	**420**	**540**	**660**	**780**	**900**	**1020**	**1140**
First-order privileged frequencies	30	90	150	210	270	330	390	450	510	570
Second-order privileged frequencies	20	**60**	100	140	**180**	220	260	300	340	380
Third-order privileged frequencies	15	45	75	105	135	165	195	225	255	285
Fourth-order privileged frequencies	12	36	**60**	84	108	132	156	**180**	204	228

Privileged frequencies for a pipe closed at one end, approximately 1.25 metres long. Only the first ten modes (odd harmonics) are used and true harmonics are in bold type. Note that no even harmonics occur among the privileged frequencies

nised for many years. Professor Benade among others has recently drawn attention to its importance and also introduced the notion of 'recipes' which I shall use in the next section.

LONG THIN INSTRUMENTS

Now before we move on to look at real instruments there is yet another complication to be considered. Suppose the pipe is conical and starts from a very small bore at one end and opens out to quite a wide bore at the open end. Suppose by some means we can start a compression from the *narrow end*; the pipe will behave just as our pipe open at both ends until the rarefaction has returned to the start. Now, because the pipe has shrunk to very small bore, the speed of the wave slows down and no real reflection occurs. Whether the end is open or closed, the wave is effectively damped out. The result is that we need only consider one journey out and one back regardless of whether the pipe is open or closed at the narrow end. The conical pipe will behave something like a pipe open at both ends as far as its modes are concerned even if the small end is closed, and it will have a full series of harmonics.

Now we can look at some real instruments. The flute and the recorder both behave like pipes open at both ends and so do most of the 'whistle' type of instrument. Organ pipes excited by edge tones with their opposite end open also behave this way and hence can contain any harmonics, depending on how the edge tone is produced; they overblow at the octave, i.e. if one blows much harder to create a higher frequency edge tone the next mode produced is the second harmonic.

An organ pipe with a cork in the end (e.g. the type known as 'stopped' or 'flute' diapason), however, will overblow at the third harmonic and also it will only respond to the odd harmonics present in the edge tone.

The clarinet has a pipe which is cylindrical except for a very short bell at the end, and a careful examination of the behaviour of the reed shows that it behaves like a closed end and reflects compressions as compressions. Consequently again the odd harmonics will tend to be reinforced more than the even ones and the clarinet overblows at an octave plus a fifth.

Now we come to the oboe. The oboe has a conical pipe and the reed is placed at the smallest part of the bore where, as we saw above, it does not really matter whether the pipe is closed or open. Thus all harmonics can be reinforced and the oboe overblows at the octave. 3.13 shows

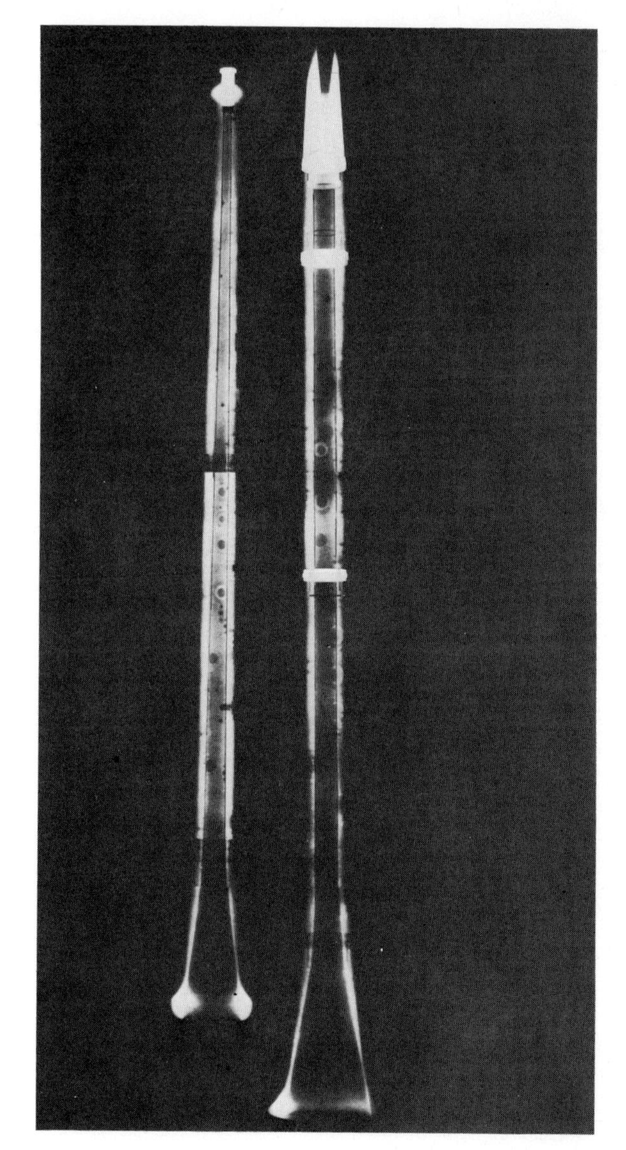

3.13 *X-ray photographs of (on the left) an oboe and (on the right) a clarinet from which the keys have been removed*

X-ray photographs of an oboe and clarinet from which the keys have been removed and one can see very clearly the form of the bore.

If you have read most of the earlier parts of this book it will come as no surprise to you to find out, once again, that we cannot deal with black and white but must take note of the greys; we cannot put all instruments clearly into the cylindrical or conical category and many of the most interesting ones are a bit of both. The brass family are splendid examples and many young physicists – equipped with the basic notion of the difference in mode patterns between cylindrical and conical pipes – have been frustrated to find that the brass instruments do not seem to fit. The privileged frequencies involved at the end of the last section provide the clue that resolves the mystery – or at least makes the solution seem possible. The lips of the player are very much more sensitive than the mechanical reeds of the clarinet and oboe families and – especially if one does the experiment with the long tube mentioned in the last section with a large mouthpiece – even inexperienced experimenters can feel their lips being 'pulled' or persuaded to vibrate at any of these special frequencies, although the only ones at which high amplitudes can be produced are the true harmonic ones.

If you are sufficiently interested and have the necessary patience you might like to share the following discussion about the notes of brass instruments with me. If not, I suggest you skip over a couple of paragraphs. To see on the one hand how complicated the brass instruments can be and on the other how some of the apparent anomalies can be sorted out let us consider a comparison between the set of notes produced by a completely conical pipe and by a completely cylindrical one closed at one end, using first of all only the true harmonics and lengths as used for *tables 3.II* and *3.III*.

Notice first of all that the frequency difference between successive modes on either pipe is the same even though, musically speaking, the intervals are different. Suppose then we shorten the cylindrical pipe a bit to make the fifth partial in tune with the fifth partial of the conical pipe. The result is shown in *table 3.IV*.

Though by no means identical the notes, except for the bottom one, are much more nearly alike and it may not stretch credulity too much to believe that pipes which are partly cylindrical and partly conical can be made to give sequences of notes which are in tune. Notice also that if we look at privileged frequencies, the second order from the third partial of a cylindrical pipe ($333/3 = 111$), the third order from the fourth partial ($466/4 = 116$) and the fourth order from the fifth partial ($600/4 = 120$), all give a privileged note round about the fundamental of the conical pipe! There are many more factors that ought to be considered, but I feel sure that we have already gone far enough to show that somewhere about here mathematical prediction becomes extremely difficult and tedious but that there is plenty of room for the craftsman to find ingenious 'rule of thumb' solutions which work beautifully in practice.

We shall now return to the main argument. Just as we cannot divide instruments neatly into those which are entirely conical and those which are entirely cylindrical, it will probably now be clear that we cannot divide them into those which are excited in a single mode by a sinusoidal edge tone and those for which all modes are excited by a regular sequence of very sharp pulses. In practice neither of these extremes occurs; no edge tone is as pure, for example, as the wave form of 1.15 c and no reed produces pulses as regularly and well defined as those in 1.15 b. There will thus be a characteristic pattern of modes within the air columns which will depend

TABLE 3.IV

Conical pipe		120		240		360		480		600		720		840	
Cylindrical pipe closed at one end	60		180		300		420		540		660		780		900

Conical pipe	120	240	360	480	600	720	840
Cylindrical pipe closed at one end	$66\frac{2}{3}$	200	$333\frac{1}{3}$	$466\frac{2}{3}$	600	$733\frac{1}{3}$	$866\frac{2}{3}$

on the shape of the pipe, the form of reed or edge and the way in which the player is blowing and controlling the tension in his lips or on the reed, and on many other factors. It should also be clear that there is no reason why the mixture should be the same for different notes from the same instrument; indeed, bearing in mind that the ratio of tube thickness to length, the velocity of the air stream and many other factors change as one moves up the scale, it would be surprising if the patterns were the same. Professor Benade describes the mixture of modes present in an air column as the 'recipe'. 3.14 shows the wave forms picked up with a microphone some distance away from a clarinet sounding three notes an octave apart to illustrate the change

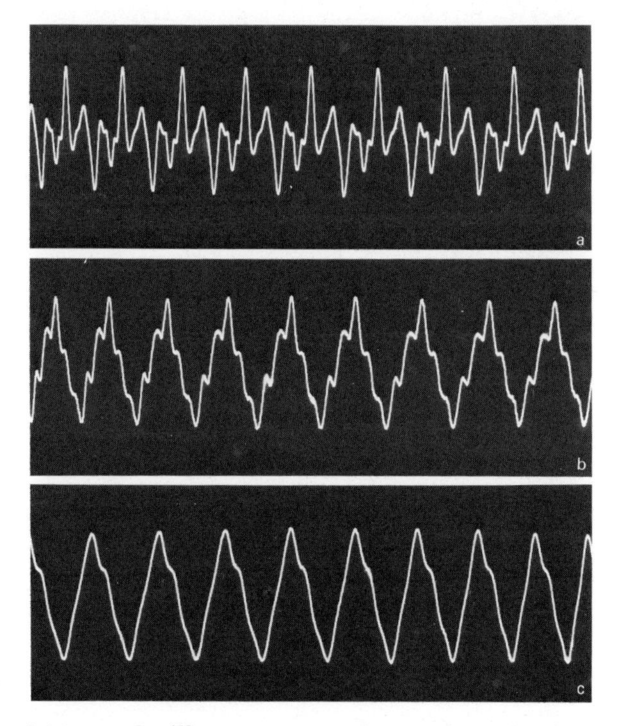

3.14 a–c *Oscilloscope traces of notes from the three registers of a B♭ clarinet: (a) played as G but sounding F (174.5 Hz) in the Chalumeau register; (b) played as G but sounding F (349 Hz) in the Clarino register; (c) played as G but sounding F (698 Hz) in the High register. The tape speed of (b) was halved on play-back and of (c) was a quarter of the recording speed so that the repeat distances of all three appear to be the same*

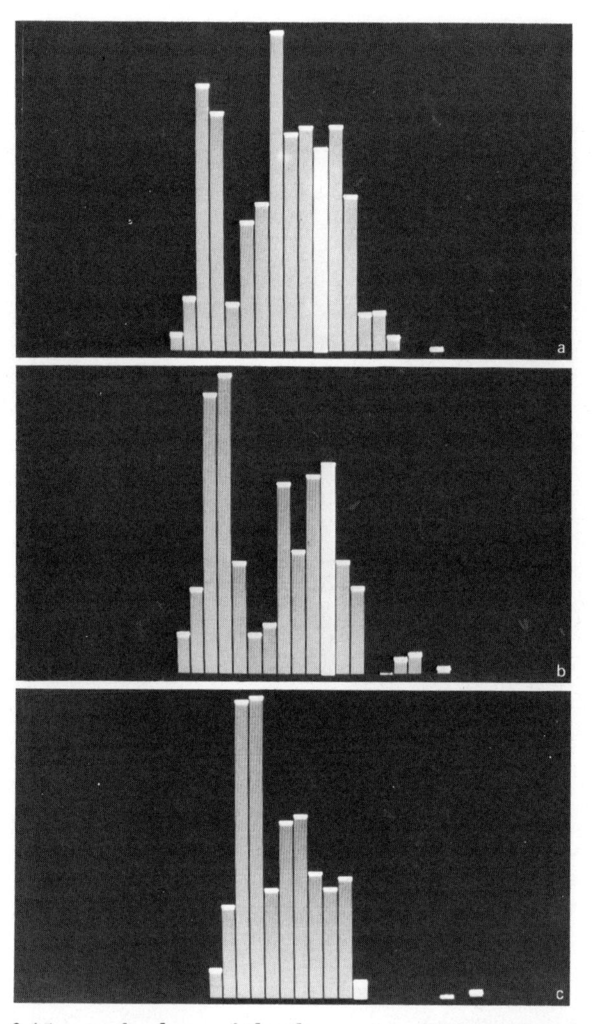

3.15 a–c *Analyses of the three notes of 3.14 replayed at the same pitch. The bright strip is at 1 kHz*

from the lowest (so-called chalumeau) register, the middle (or clarino) register and the high register. The notes are played as G on a B♭ clarinet and because of the mysteries of transposition sound F, 174.5 Hz, 349 Hz and 698 Hz. 3.15 shows the corresponding frequency analyses and the markedly different recipes are clearly seen. A convincing way of demonstrating this aurally is to record all three notes and then to play back the high one with the tape running at quarter speed, the middle one at half speed and the low one at its true speed. The notes then sound to have the same pitch and their differences in quality are obvious. The same technique was used in producing 3.14, so that the repeat distances of the wave forms are the same.

3.16 shows the wave forms and recipes for four brass instruments – a euphonium, a baritone horn, a tenor trombone and a piece of brass tubing just over 8 feet long. All are the same length – but the recipes are clearly very different and we need to think in more detail about the origin of the differences – though the clues have nearly all been given already. One major point

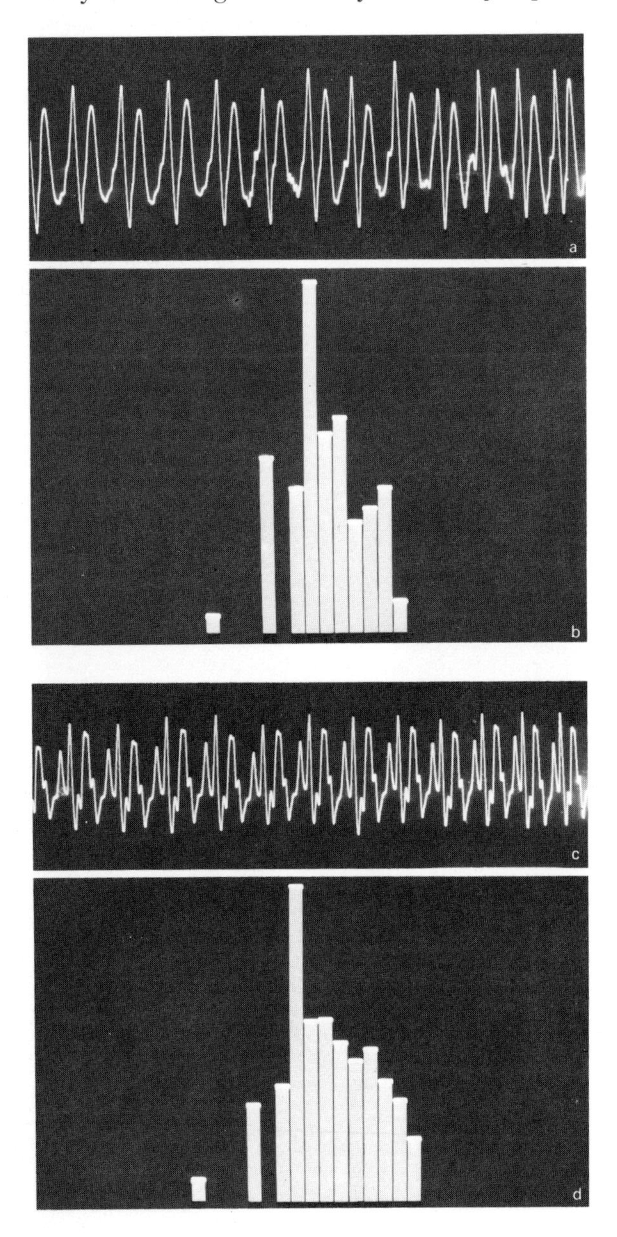

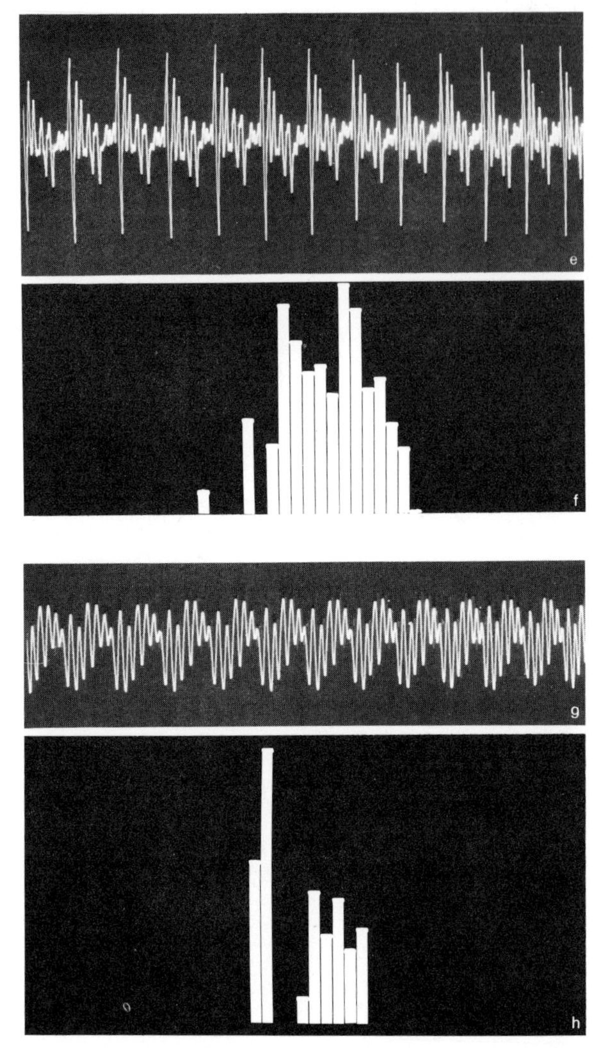

3.16 a–h *Traces and analyses of the same note (B♭, 116.5 Hz) from (a), (b) euphonium; (c), (d) baritone horn; (e), (f) tenor trombone; (g), (h) 8-foot-long brass tube*

is that we are beginning to depart from 'long thin instruments', as can be seen from 3.17. Apart from the fact that the three brass instruments are to some extent coiled up, the euphonium is obviously quite fat at the end and could not by any means be described as long and thin in the sense that the long brass tube could. We must now, therefore, consider vibrations in systems which are not long and thin.

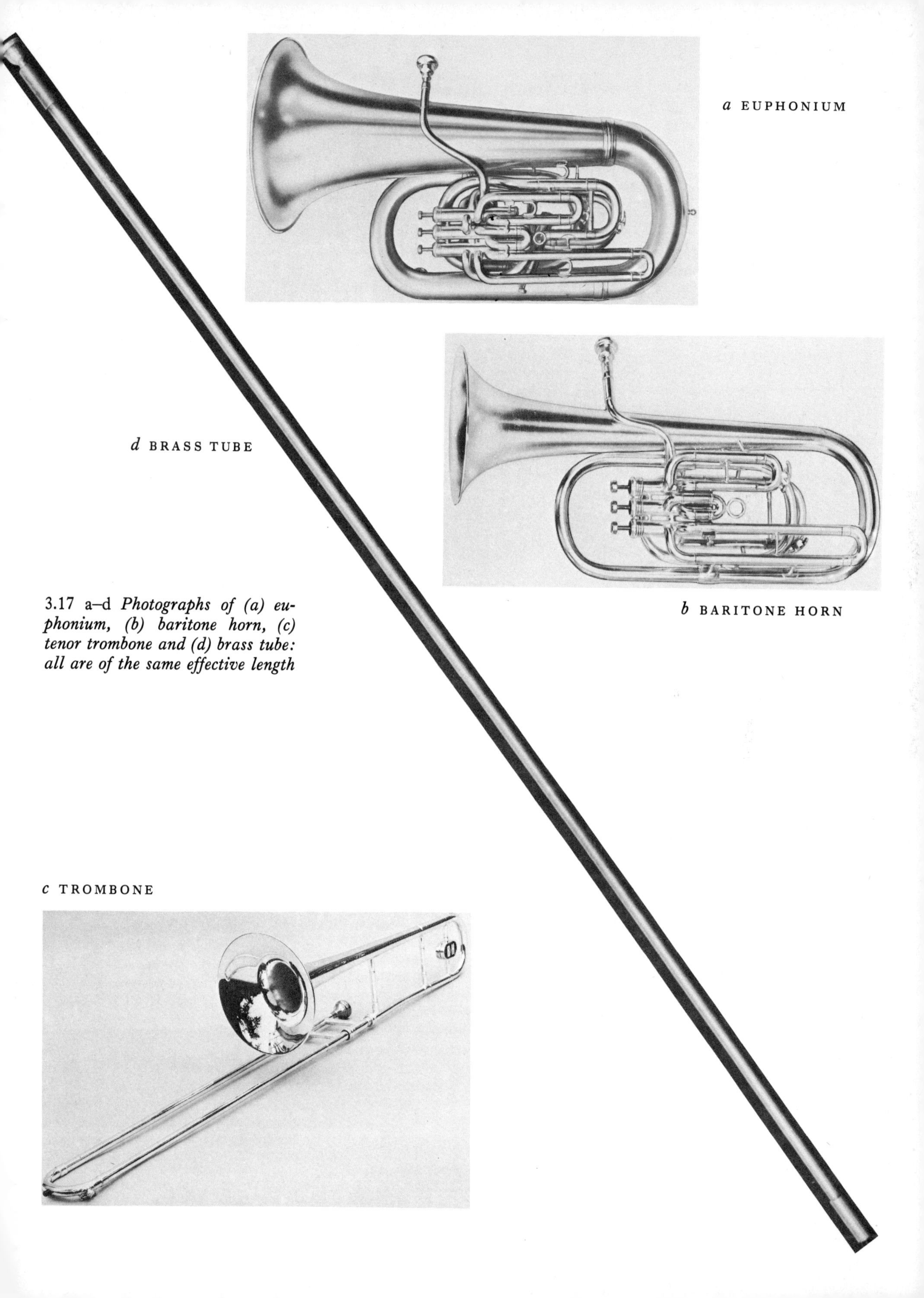

3.17 a–d *Photographs of (a) euphonium, (b) baritone horn, (c) tenor trombone and (d) brass tube: all are of the same effective length*

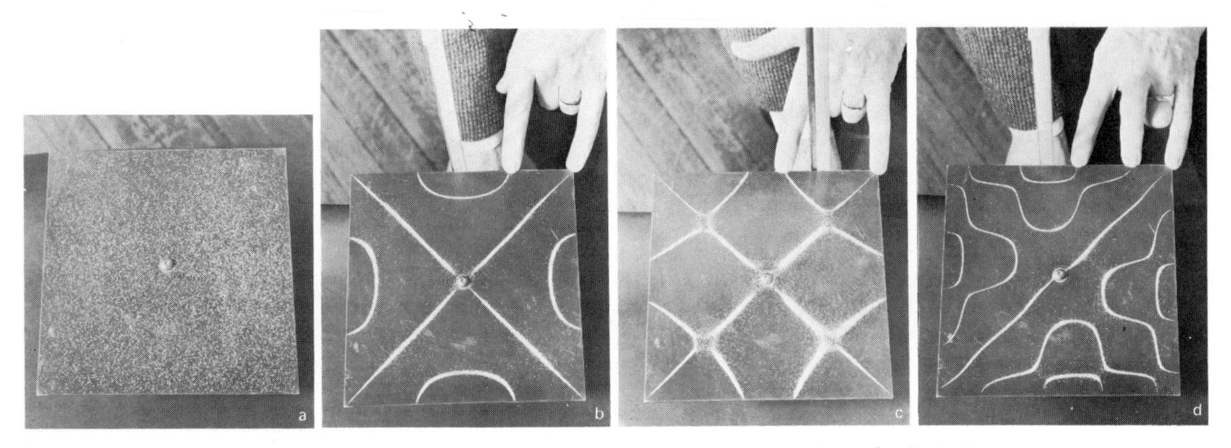

3.18 a–d *Bowed Chladni plate. The frequencies are (b) 1200 Hz; (c) 880 Hz; (d) 2760 Hz*

RESONANCE IN PLATES AND BOXES

Although we have considered long thin things for simplicity and because they do dominate the western orchestra, nevertheless vibrations may travel, be reflected and probably build up resonances in objects of any other shape; it will equally be obvious that the results will be a good deal more complicated. Let us start by thinking back to the rubber rope of 1.25. The rope has one dimension very large and two dimensions very small; we shall now make two dimensions large and one small to give a plate. The modes of vibration of a plate can be demonstrated by an elegant experiment due to Chladni (1787). In its original form a hard brass or steel plate clamped firmly at its mid point can be made to emit a great many different– but quite precise – notes by bowing the edge at different places and holding the edge at others. Sand scattered on the plate dances as the plate vibrates and tends to congregate near to the lines where the vibration is least. These lines correspond to the nodes, or points of no displacement on the rope, in say 1.25 d and are called nodal lines. 3.18 shows three such modes being excited in the classical way. A simpler alternative, demanding no skill at all on the part of the experimenter, is to replace the centre clamp by a vibrator unit – in effect a large loudspeaker unit with no cone attached – which will oscillate in step with any electrical signal fed to it. By feeding pure tones from an electronic signal generator the plate can be set in vibration. Starting at a low frequency and gliding slowly upwards we find a sequence of specific points at which the plate gives out a much louder note. These are the resonances or resonant modes of the plate. Two series of photographs of sand patterns on differently shaped plates are given in 3.19 and 3.20. In each case the driving frequency is given and it will be obvious that frequencies do not fall into a harmonic series (unless one considers them as harmonics of a fundamental of 5 Hz! – this is scientifically correct but clearly makes musical nonsense). It turns out on more detailed analysis that as the shape of the plate is made more complex, fewer very strong modes occur – though a large number of weak ones may be detected audibly as increases in loudness that are not sufficiently well defined to produce a sand pattern. If one tries to follow the kind of reasoning we applied in thinking of the build-up in pipes one can see that this result is not surprising. In a square or circular plate we should expect that reflections from the edges would all coincide and respond to the same driving frequency; in an asymmetric plate the likelihood of finding many frequencies for which reinforcement occurs for several waves in different directions is clearly much smaller.

The same sort of argument may be extended to three-dimensional objects such as boxes or vessels of any shape and again it will not come as a surprise to find that a cube or a sphere

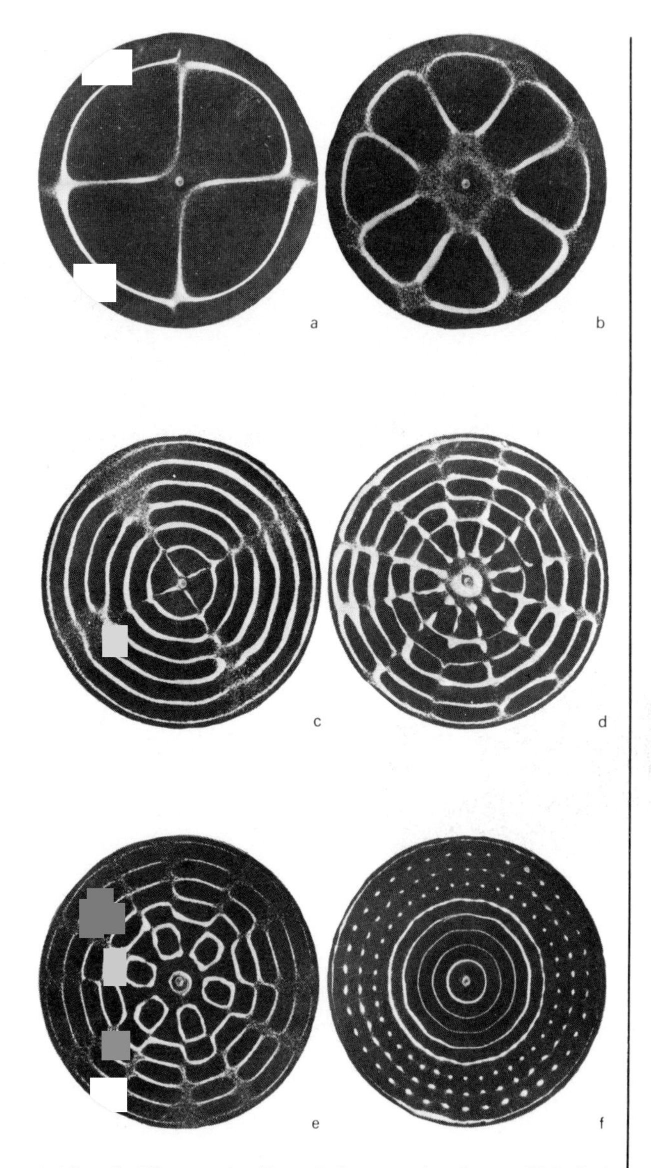

3.19 a–f *Electronically driven circular Chladni plate at frequencies (a) 280 Hz; (b) 865 Hz; (c) 4500 Hz; (d) 7500 Hz; (e) 9500 Hz; (f) 10,000 Hz*

provides more clearly defined modes than more complex shapes. The magic wand mentioned earlier is a useful way of revealing three-dimensional mode patterns and in 3.21 we see time exposures of the wand being moved about inside a perspex cube being excited to resonance by the sound from a small loudspeaker placed at a hole in the base of the vessel. Though the movement of the wand is necessarily some-

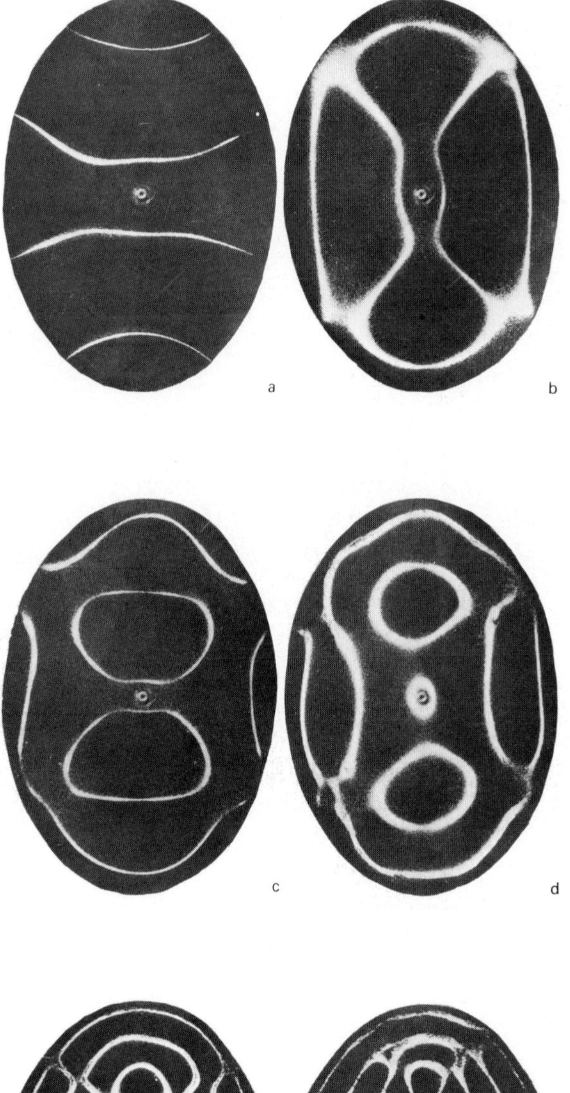

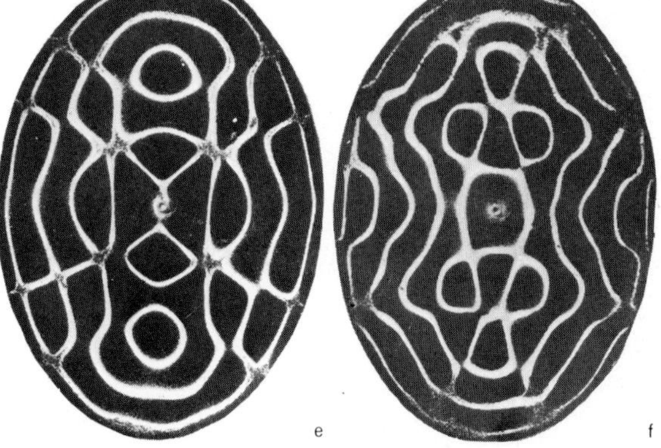

3.20 a–f *Electronically driven elliptical Chladni plate at frequencies (a) 345 Hz; (b) 650 Hz; (c) 950 Hz; (d) 1500 Hz; (e) 5000 Hz; (f) 6750 Hz*

3.21 a and b *Time exposure of Perspex cube with 'magic wand' moved around; the bulb lights at places where the pressure variations are a maximum and is extinguished at pressure nodes: the loudspeaker is fed with tones (a) at about 200 Hz; (b) at about 1600 Hz*

what erratic the modal pattern is nevertheless quite clear. We shall refer back to these three-dimensional modal patterns – though on a much larger scale – in chapter five when we discuss the way in which sound behaves in large rooms or concert halls; at the moment we are more concerned with the fact that vessels tend to have only one or two very strong resonances and a great many weaker ones. Clearly then, if we are to use boxes or bottles as amplifiers to make the sounds of instruments louder they will certainly behave like spectrum shapers and will impose a formant characteristic on the sound fed in. Clearly also, if we use simple regular shapes we shall find one or two sharp resonances whereas complicated shapes are more likely to have a wide range of smaller ones.

Helmholtz (1877) used a series of spherical vessels as frequency analysers. Each has a large hole in one side to pick up the sound and a narrow tapered tube at the other side to fit the ear of the observer. 3.22 shows a set made of brass from the Royal Institution collection.

Such resonators are surprisingly effective in picking out a specific frequency from a complex sound. The effect is, in fact, familiar to most of us in the childhood myth that, if you hold a large cowrie or whelk shell to your ear, you can hear the sea. What is happening, of course, is that the shell, because of its complex shape, is picking out of the general noise background a great many frequencies and amplifying them; as the background sound varies so the sound amplified will vary. The same effect can be demonstrated by cupping your hands over your ears; make sure that you keep your thumbs in contact with your head behind your ears and then vary the opening between the front edge of your hand and your head and alter the degree of cupping of the palms. In almost any surrounding there is enough background noise to do this experiment; if your surroundings happen to be exceptionally quiet turn up the volume on a small radio not tuned to a station to produce a crude approximation of white noise.

The most important family of instruments which makes use of amplifying properties of plates and boxes is the string family. We include not only the fiddles, but also guitars, mandolins, lutes, etc., and harps, harpsichords, virginals, pianos, etc. Our problem is that we must amplify the sound of the strings which, by themselves, are much too weak, but that if we use resonances of plates or boxes to get the required result we run the risk of changing the quality of sound enormously in the process. The trick, as has been pointed out before, is to make this a positive advantage in enhancing the parts we need and a great deal more about how this is achieved will be said in the next chapter. It is obvious, however, that if we produced a violin with a very strong sharp resonance, at some one particular note in the middle of its range and not much response anywhere else it would not be a very useful instrument. In the same way an amplifier to be used with a record player which had a response characteristic (or formant characteristic) like that of (say) 3.3 e, would change the quality of sound in what might be a very objectionable way – though here we are once more jumping the gun and trespassing on chapter six, where we will consider the effect of the waves produced by such systems on our ears.

A very elegant experiment to demonstrate both the amplifying properties and the formant characteristics of several instruments was performed at the Royal Institution in the 1860s by Professor John Tyndall, though he attributes the idea to Sir Charles Wheatstone. A wooden rod was passed through a hole in the floor of the lecture theatre and through the floor below so that its lower end could rest on the sound board of a piano; the upper end projected two or three feet above the floor of the lecture theatre. The rod was held in place by padding so that it did not touch the sides of the holes and to eliminate air transmission through the holes themselves. A pianist performed in the basement and the level of sound transmitted up the wooden rod was too low to be heard in the theatre. If, however, the body of a harp, 'cello or violin was placed in contact with the top of the rod not only could the piano be heard clearly but the sound quality changed according to the instrument used for amplification, a very beautiful demonstration of the formant. Tyndall was a great enthusiast and expended enormous effort in getting his demonstrations to work well and in describing them. His own description of the climax of this experiment in *On Sound* (1867) is worth reproducing:

3.22 *Set of brass Helmholtz resonators from the collection at the Royal Institution*

'What a curious transference of action is here presented to the mind! At the command of the musician's will, the fingers strike the keys; the hammers strike the strings, by which the rude mechanical shock is converted into tremors. The vibrations are communicated to the sound-board of the piano. Upon that board rests the end of the deal rod, thinned off to a sharp edge to make it fit more easily between the wires. Through the edge, and afterwards along the rod, are poured with unfailing precision the entangled pulsations produced by the shocks of those ten agile fingers. To the sound-board of the harp before you the rod faithfully delivers up the vibrations of which it is the vehicle. This second sound-board transfers the motion to the air, carving it and chasing it into forms so transcendently complicated that confusion alone could be anticipated from the shock and jostle of the sonorous waves. But the marvellous human ear accepts every feature of the motion, and all the strife and struggle and confusion melt finally into music upon the brain.'

I have already mentioned the way in which the Royal Institution preserves all kinds of instruments and apparatus from the past; they have also preserved the hole in the floor that

Tyndall made, and the present author was able to reproduce the experiment with the added advantage of a television camera in the basement to show that the piano was actually being played when no sound could be heard in the theatre.

We have now reached a point at which we can be little more specific about terminology. So far we have talked about formants and frequency response characteristics as if they are interchangeable terms, and indeed to some extent they are. The term 'formant', however, does have rather special connotations and is usually used when a particular characteristic is imposed on a variety of sounds. For example, if you were to place a Helmholtz resonator in your ear and listen to a piece of speech or music it would be correct to describe the effect of the resonances as a formant imposed on the sound. The shape of a bugle – which has no keys or slider – remains fixed and the only way in which the player can change the note is by varying the excitation of the lips and breath – hence one would expect to find a formant characteristic here. Though again we must be careful because it could be – as we shall see in the next section – that the behaviour of the sound waves at the bell end may be different at different basic frequencies. We might expect a fixed formant in a harp, piano, harpsichord, etc., and to a large extent in the other strings. There is a slight complication with the instruments which have strings whose length is varied by the player, however, in that the way in which the strings are coupled to the body varies and so does the pressure on the bridge and these may change the characteristics slightly over the range. In the woodwinds the whole shape is changed as the notes are changed and there is therefore more likelihood that the characteristics of amplification will change. We are now very close to the subject of the next section – getting the sound into the air – but there is just one more point to make before we move on.

In chapter one we looked at the vibration of strings with unequal weights placed on them. It might be helpful now to look briefly at the vibrations of tubes with unequal bulges. 3.23 shows three glass devices which might look more in

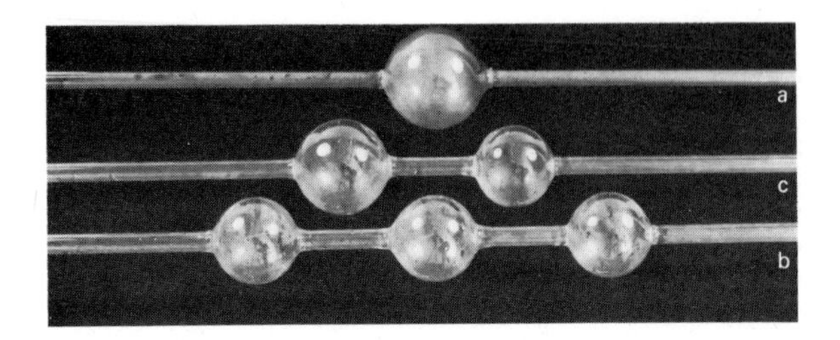

3.23 *The glass vessels, each of which is the same length and has the same volume*

place in a chemistry lab than in an acoustics lab. All have the same volume and the same length. To demonstrate the point being made the thumb is placed over the lower end to seal it and air is blown across the other end to produce a note– as in the pan pipes. The frequencies produced are (a) 471 Hz, (b) 660 Hz and (c) 570 Hz. Now all three tubes are turned upside down and the experiment is repeated. Again (a) gives 471 Hz and (b) gives 660 Hz, but now (c) gives 535 Hz. What does this tell us? Simply that the response does not solely depend on length or on volume but also on distribution of volume or shape. Of even greater importance however is the pattern of modes that are produced. If (a), (b) and (c) are blown harder they do not all produce overtones with the same frequency ratio to the fundamental. Tube (a) overblows at 942 Hz – exactly the octave in spite of the lower end being closed! Tube (b) overblows at 1292 Hz – a very flattened octave. (c) behaves like (a) when the larger bulb is near the blown end and overblows at 1140, but when reversed it overblows at 1035 Hz – again a very flattened octave. This means that the recipe of the vibration complex produced inside will be different. Indeed it turns out that by changing the three variables of overall length, overall volume and shape we can produce a series of vessels which produce the same set of notes but with quite different vibration recipes or tonal qualities. This, of course, is precisely the point we had reached at the end of the last section in discussing the euphonium, baritone horn, tenor trombone and the simple length of tubing. When we go on to consider the problems of playing all the wide range of notes demanded of a modern orchestral brass instrument in the next chapter we shall meet further ramifications of these three variables. The other point that emerges very strongly from the experiment with the four brass instruments is that the notes from the plain tubing are very dull and muffled and do not have anything like the carrying power of the other three. A first guess might be that the flare at the end has something to do with the result and, as we shall now see, it would be right.

GETTING THE SOUND OUT

In chapter two we talked about the tuning fork and the point was made that its prongs cannot produce very big waves in the air unless coupled to a table or other large plate. In the same way our 8 feet of brass tubing may have very violent vibrations going on inside it, but only the end of the tube is in contact with the open air and it is not really surprising to find that the euphonium with its large bell produces a very much louder sound. The question we must look at in this section is not just how the sound gets out of various instruments, but whether the way it gets out has an effect on the quality. Does it provide yet another formant? A simple experiment may help here. An electronically produced sound complex is fed to a small loudspeaker in a box and the flare of a trumpet or of a trombone can be fitted to the hole in the box. 3.24 shows the arrangement; (a) and (b) show the wave form and spectrum of the sound coming out of the box with no horn; (c) and (d) show the wave form and spectrum from the trumpet horn; and (e) and (f) the wave form and spectrum from the trombone horn. The quality difference is very noticeable to the ear; both increase the volume considerably and the trumpet horn makes a much more harsh and reedy sound than the trombone.

What about the woodwinds, however? Does the bell on the clarinet or oboe have a similar effect to the brass horns? The simple straight answer is that it does not. In order to understand why, we must ask ourselves how the sound emerges from woodwind instruments. The speediest way of answering this question is to take a clarinet, remove the bell, play the low-register note obtained by covering the thumb and three finger holes with the left hand (C, which sounds B♭ on a B♭ clarinet) and alternately close and open the bottom end of the instrument where the horn usually fits with the palm of the right hand. The hand makes virtually no difference! A slightly less effective but possibly more accessible experiment is to play a recorder with the thumb and three fingers of the left hand in position and again to close and open alternately the bottom end with the palm of the right hand. Again there

3.24 *Trumpet and trombone horns fitted to a small loudspeaker unit*

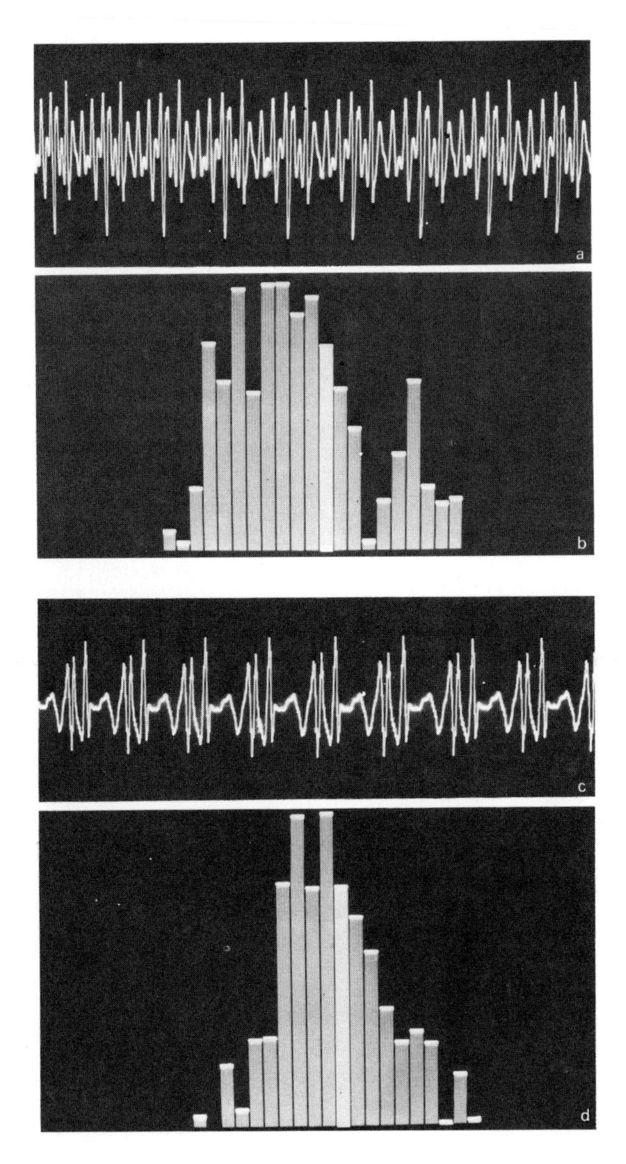

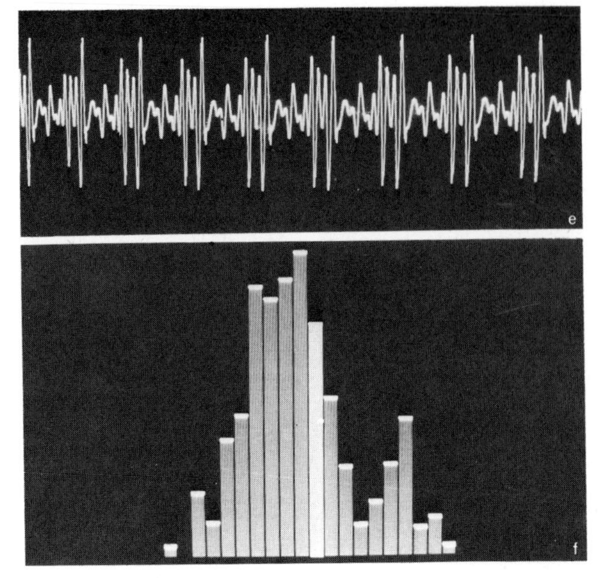

3.24 a–f *The wave forms and spectra are as follows (in the spectra the bright band is at 1 kHz): (a), (b) sound emerging from speaker with no horn; (c), (d) sound emerging from trumpet horn; (e), (f) sound emerging from trombone horn*

is virtually no difference and we realise that the sound must be emerging from the holes in the side wall.

3.25 shows the wave form as detected by a so-called probe microphone at various parts of a clarinet while the same note (D – 294 Hz, fingered with thumb hole and top finger hole closed) is being produced. The probe microphone is merely a small pressure-sensitive microphone which has a thin flexible tube attached that can be moved around or inserted at various points in an instrument. For 3.25 a it was actually attached to a hypodermic needle which was inserted through the wall of the mouthpiece (an awful thing to do to a clarinet – but in a good cause!). For 3.25 b the probe is inserted into the first open hole of the clarinet, which is the second hole down. In 3.25 c the probe is in the fourth hole from the top and in 3.25 d it is inside the bell at the end of the instrument. A similar experiment with an oboe sounding B (= 494 Hz) gives the trace of 3.25 e when the probe is in the top open hole and 3.25 f when inside the bell at the end. It is clear that the sound emerging from the bell is not only weak, but also has only the higher-frequency components. Why then do woodwinds have a bell at all? It is of course needed to compensate for the fact that there are no holes from which the sound can emerge other than the one at the end when the very lowest one or two notes are played. The bell helps to 'match' the vibration in the pipe to the air outside for these one or two notes only and to avoid an abrupt change in quality when the bottom few notes are used.

The interesting point that emerges is that the recipe inside the pipe is not necessarily the same as that which is radiated from the holes. So we have yet another formant and we see that the finger holes are not there just to permit us to play tunes but also to get the sound out and further modify its quality. The relative positions of the hole affect their formant properties – they both are, broadly speaking, rather like the set of rods on a television aerial and everyone knows that different spacings are needed for different frequencies. The position of the holes relative to the patterns of compressions and rarefactions inside the tube for the different modes and for different notes is obviously the critical point. But how can the instrument maker get the holes in the right place both to permit playing in tune and to produce the right 'radiation formant'? We are beginning to trespass on the technology of the next chapter – but the solution lies in the fact that the *diameter* of the holes can be changed as well as their position and this gives the manufacturer the additional control he needs.

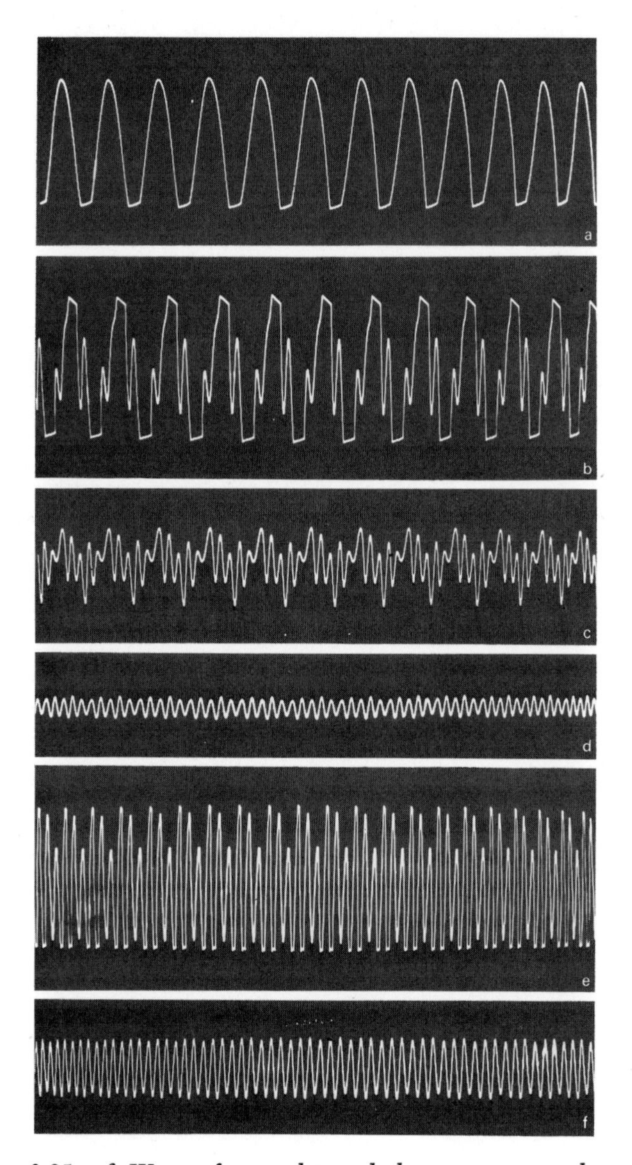

3.25 a–f *Wave forms detected by pressure probe microphone: (a) in the mouthpiece of a clarinet sounding D (294 Hz); (b) as (a) but with the probe tube in the first open side hole; (c) as (b) but with the probe tube in the next but one open hole; (d) as (c) but with the probe tube in the open bell; (e) in the first open hole of an oboe sounding B (494 Hz); (f) as (e) but in the open bell*

Professor Benade describes an experiment that is very helpful in understanding these points. It involves playing a note with a clarinet mouthpiece fixed to a 1-foot length of metal or plastic tubing – such as electrical conduit – and comparing its quality with that from a 2-foot length which has several finger holes bored in the lower half so that it plays the same note as the first. The first sounds dull and muffled, but the second tube displays some of the richness of clarinet tone if the size and spacings of the holes are something like that on a clarinet; if the spacing and size of the holes are changed it is possible to find combinations for which the sound is even weaker and more muffled than for the plain tube.

I propose now to plunge you into very deep water. You may prefer to skip on to the next section, but the very exciting and elegant ideas that are emerging from Professor Benade's laboratories are too good to miss out, and I want to try to include at least a general impression (which I hope is a fair, though much simplified, view) of them.

We have already likened the regular spacing of the holes on a woodwind to the rods on a television aerial and it turns out that, just as one needs to change the size and spacing of the rods for different frequency bands, so the size and spacing of the holes has a bearing on the response frequencies of the pipe. If a wave travels down a pipe with no holes and then meets a section in which holes are regularly spaced the lower-frequency components will be reflected back at the first hole just as though it was an open end (as the simple theory would predict), but the higher frequencies 'leak' past and travel down to the end, more or less ignoring the open holes. The frequency at which the change-over – or cut-off frequency as it is called – takes place depends on the size and spacing of the holes in relation to the bore of the pipe and the thickness of its walls. This explains why in 3.25 the waves emerging from the bell have mainly high-frequency components and those from the first side holes have low-frequency components. It turns out, too, that the flared parts of brass instruments also introduce cut-off frequencies and hence clearly affect the recipe that can be built up inside the pipe and that which is radiated from the horn. Further, when the French horn player places his right hand inside the bell in the normal playing position his hand actually modifies the cut-off frequency and hence enables him to alter the tone and playing quality of the instrument. Mutes in the end of brass instruments behave similarly.

Now there is another complication! When one delves into the real facts about the behaviour of pipes which are not plain cylinders or cones, but have holes in the side or 'bumps' resulting from holes which are closed by keys or the fingers, it turns out that all the overtones are not necessarily strictly harmonic. Though this might be expected to affect the quality of tone, one would not, at first sight, think that it would have any more profound effects – but because of our non-linearities, already cited as so important, it turns out that they do.

Let us think about any one mode of vibrations; the variations in pressure along the pipe are sinusoidal, but when they impinge on the reed after the first reflection from the open end or from open holes the response of the reed may not be sinusoidal. 3.26 shows how the reed is likely to move in response to various forces on it. The horizontal axis of the graph shows the

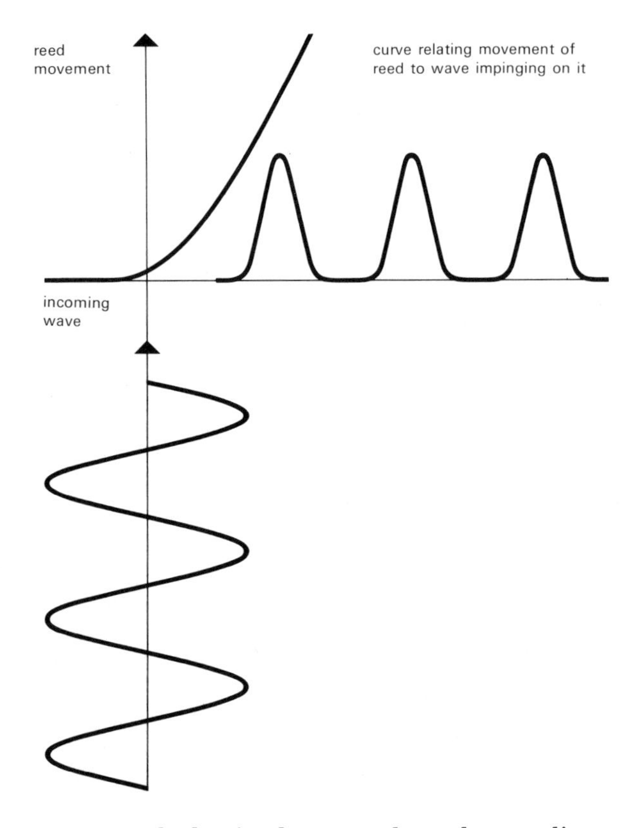

3.26 *Graph showing how a reed or other non-linear device might respond when fed with a pure sine wave. The input is shown graphically coming from below and the output going out to the right*

compression or rarefaction in the air impinging on the reed and the vertical axis shows how it actually reacts. To the left of the centre vertical line the reed has closed and so no more movement occurs whatever the incoming wave does. The result of a sinusoidal wave (shown coming up from below) is thus to produce a series of pulses (shown moving off to the right). This is a simplified view of the effect of the non-linearity of the reed.

These pulses (following once more the pendulum argument and 2.25) can excite many different modes, but *only if they are strictly harmonic* will the reed be able to keep all of them going. If they are not harmonic then one or other mode will predominate and the others will get out of step, and the resulting lack of cooperation may even stop the pipe from making a sound altogether. This is a further reason why the precise positioning of the holes is so important.

One can improve the tone by bringing the modes into tune to make them strictly harmonic. Two- or three-foot lengths of rubber hose with a clarinet mouthpiece and reed at one end may be used to demonstrate this. 'Squashing' the tube at various places has different effects in different modes; at a pressure node it sharpens the pitch and at an anti-node it flattens. By experimentally squashing the pipe at various places one can find a point which makes the fundamental and the first overtone accurately in a 1:3 frequency ratio; when the fundamental is then played it has a much richer tone and is easier to produce than when the two are out of tune.

Professor Benade has developed the art of encouraging cooperation between the modes in a clarinet for each individual note to such a point that he can produce exciting improvement in both tone and ease of playing even in already good instruments. I should hasten to warn, however, that the process involves immense skill, patience and experience and often the changes in dimensions required are fractions of a thousandth of an inch. There is no short cut in acquiring the necessary abilities! Here we are already moving into the realms where science and craftsmanship must go hand in hand.

One of the problems, of course, is to find enough variables so that changes can be made to get the pitch of each note exactly right, to get the right cooperation between the modes, to get the right cut-off frequency and the right radiation characteristics from the side holes, all simultaneously. As a final example of a variable which the simple theories usually ignore, consider 3.27. It shows a set of five identical recorder mouthpieces that have been fitted to a set of identical plastic tubes. The first is left unchanged, a small hole is drilled in the second about 2 inches from the lower end, a larger hole is drilled at the same point in the third and even larger in the fourth, and the fifth is cut off at the position of the holes. Common sense might suggest that all but the first would give the same note, whereas in fact all five are different. The frequencies given by the five shown in the photograph are given in the caption.

You should now see just how interconnected the various aspects of reed instruments are. The cut-off frequency determines the modes that contribute, the precise and detailed shape of the bore and holes determines the frequencies of the modes, the non-linearity of the reed links them together and only if their frequencies are right do they cooperate.

FORMANTS IN SPEECH

The human voice displays many of the characteristics of the instruments that have been discussed in this chapter. It involves a heterophonic reed

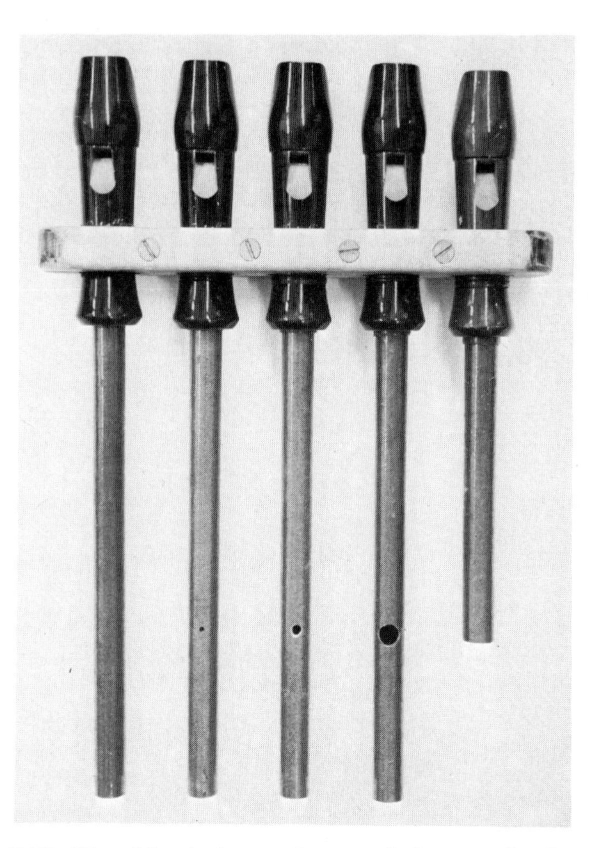

3.27 *Five identical recorder mouthpieces and tubes differing only in the size of the single hole pierced in the side or in the length. The corresponding frequencies are: 538, 587, 622, 668 and 698 Hz respectively*

which has some idiophonic qualities as the basic source! The vocal chords can be tightened and slackened by muscle movements, and that combined with changes in the velocity of air forced through by the lungs provides a basic vibration and sets the pitch of the voice whether in speaking or in singing. The result is a very 'spiky' wave form resembling that produced by an oboe or clarinet reed and hence capable of exciting a wide range of frequency responses. There is also a considerable component of background white noise from the turbulence in the air – you can hear this clearly on its own if you whisper 'Ah', but it is still there together with the regular pulse wave form if you say or sing 'Ah'.

This basic combined wave then passes through a series of tubes – the larynx, etc. – to which a series of cavities – the sinuses, mouth, nostrils, etc. – are all attached and is finally radiated from an aperture of variable size and shape – the lips. All kinds of changes in the shape of these cavities are possible by muscle control, by changing the position of the tongue, by altering the position of the lips – and of course less controllably by changes due to infection or the accumulation of mucus as when one has a cold. All these changes vary the formants imposed on the basic sound and we need to identify three types of formant. First the one which is not easy to change and is characteristic of the person speaking, secondly a consciously controlled one which determines the quality of sound produced from the harsh rasp of anger to the gentle murmur of a mother soothing a child, and thirdly, again controllable, that which determines the vowel quality of speech.

Several interesting experiments can be done to demonstrate the existence of formant characteristics and to show that they may be controlled. The frequencies of the various formants are determined by the shape and sizes of the cavities in relation to the velocity of sound in the air which usually fills them. If you fill your lungs with a mixture of helium with a few per cent of oxygen (to ensure survival!) the velocity of sound is so much higher that the resonant frequencies are all raised and the voice acquires a most peculiar characteristic reminiscent of that of the cartoon characters, the Chipmunks. This is not too surprising as their voices were produced by a technique involving the speeding up of tape-recordings.

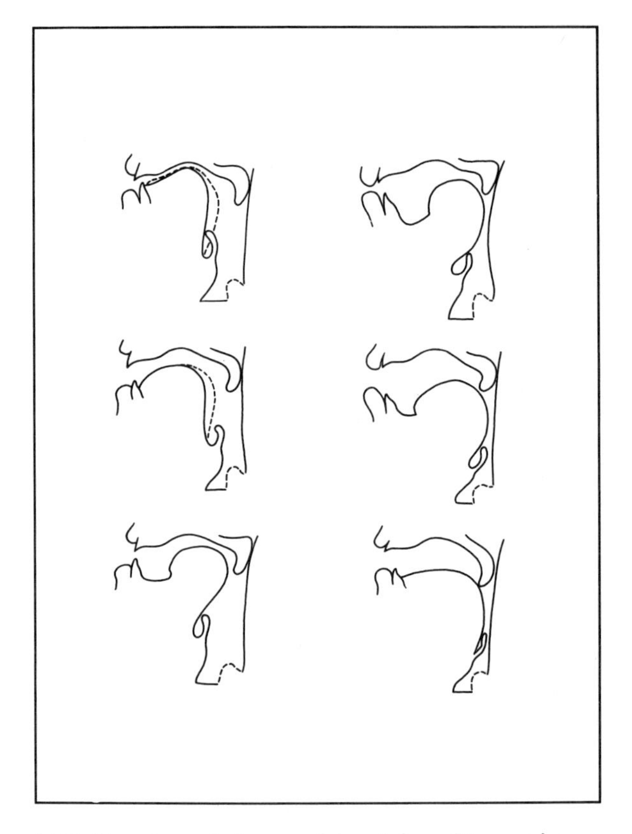

3.28 *Tracings of the vocal tract from X-ray pictures taken while six Russian vowels were being spoken*

If an electric razor is placed on the cheek with the mouth open, the quality of sound that emerges can be changed by shaping the mouth as though different vowels were to be spoken but without making any noise with the vocal chords. With practice it is possible to shape the mouth to pick out individual harmonics of the tone of the razor. From time to time variety performers play tunes by tapping their cheeks or their heads with the open mouth held close to a microphone; these and the well-known Jew's harps all use mouth and throat formants as their basic frequency control.

Studies of speech and of vowels in particular have been made for many years and some of

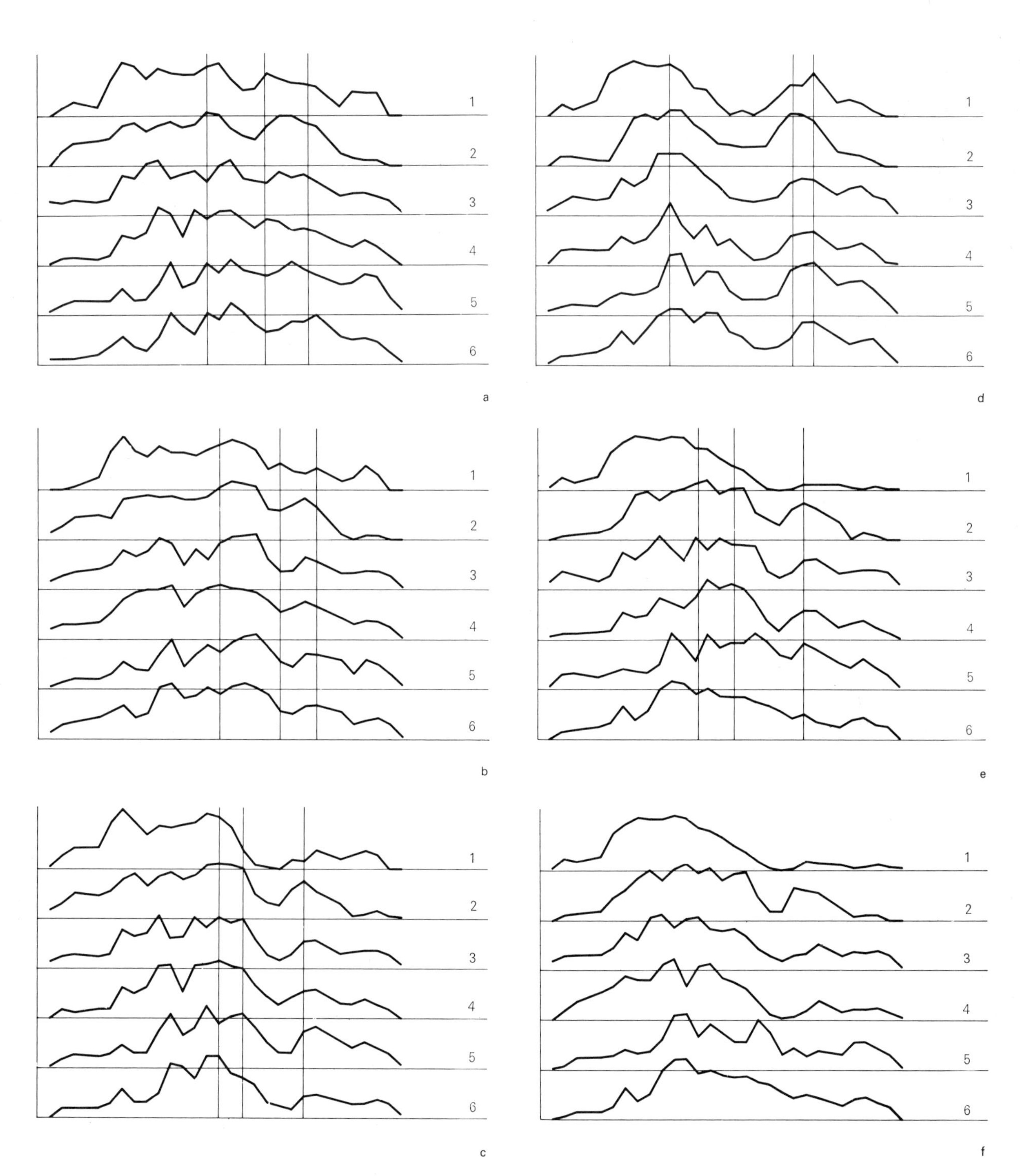

3.29 a–f *The frequency distribution of the vowels spoken by six different people (1 and 2 are men, 3 and 4 are women, 5 and 6 are children) is shown.*

The vertical lines are the generally accepted regions of the three formants appropriate to the vowels. (a) = Ah. (b) = Aw. (c) = Er. (d) = Ee. (e) = Oh. (f) = Ooh.

the earliest and highly ingenious work was done by Sir Richard Paget, who made models of the throat and mouth cavities with cardboard tubes and Plasticine and demonstrated artificial vowel sounds in the 1930s. Among the more recent and highly sophisticated work is that of Professor Gunnar Fant of the Royal Institute of Technology in Stockholm. His studies, based on X-rays of speakers while producing sounds, are classics not only in relation to the results but also for the elegance of the methods used. 3.28 from Professor Fant's book shows traces of the vocal tracts made from X-ray photographs as a series of Russian vowels is being spoken.

The current theory of vowel sounds in speech depends on a total of five factors. The first is the fundamental frequency of the vocal chords and this varies considerably between speakers. For example, in men it might be in the range say 50–250 Hz, whereas in women it might be an octave or so higher. The vocal chords, however, produce a very wide range of harmonics and these are modified by four identifiable formant peaks which change very little between men and women. 3.29 shows a series of graphs prepared to illustrate these points using the real-time frequency analyser. Six speakers were invited to take part, two men, two women and two children; the men were one from the North of England (1) and the other from Wales (2); the women were one from Wales (3) and the other from central Europe (4); the children, a boy (5) and a girl (6), came from Wales. 3.29 a, b, c, d, e and f show respectively the frequency graphs of the six speaking the vowels Ah, Aw, Er, Ee, Oh, Ooh, and the lower three of the four generally accepted formant regions F_1, F_2 and F_3 are drawn in as vertical lines. The similarity of the traces for a given vowel in the region about 1 KHz is very marked; similarly the resemblance between the traces for a given *speaker* at the low-frequency end is obvious. The increase in fundamental vocal-chord frequency as one moves from male to female and then to children's voices is also clear.

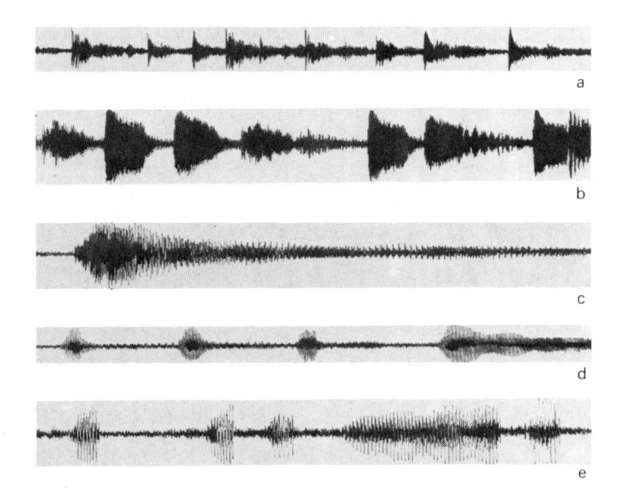

3.30 a–e *Envelopes of sounds produced by various instruments: (a) xylophone; (b) piano single notes; (c) piano chord; (d) clarinet – staccato; (e) bassoon – staccato*

WOBBLING AND FADING AWAY

Finally under the heading of 'growing and changing' we need to pay just a little attention to what happens after a note has been initiated, grown to its full size, and been modified by various formants arising from amplification. Will it then stay steady in amplitude and constant in tone? The answer in general is no. Again the factors involved are many and complex; we can only pick out one or two of the most obvious, and the first is vibrato. This is a subject which always raises problems and is certainly an area where physicists should step rather warily. The 'tremulant' stop on an organ or the 'wobble' of a third-rate soprano can drive some listeners to despair, and yet the beauty of the vibrato on the long notes from a virtuoso violinist or oboist is quite haunting and is an integral part of the expected tone quality.

In stringed instruments the effect is produced by rolling the tip of the finger, which defines the length of the string, backwards and forwards and so producing slight variations in the vibrating length and hence frequency. In reed instruments the variation is produced by changes in pressure both of the lips on the reed and of the air pressure used, and hence both amplitude and frequency can be modified. Some clues about why the effect can be pleasant will be given in chapter six; for the moment it is merely worth noting as an important factor in the development of a musical note, and that it is extremely difficult for most singers and instrumentalists to eliminate it completely even if they make a deliberate attempt to do so.

The second factor is the way in which the note actually decays. We have covered this point indirectly already, but the effect is worthy of mention in its own right. The slow-speed oscilloscope is the most useful device for demonstrating here and 3.30 shows examples of the overall envelope – the scan is too slow to show individual waves – of notes of (a) a xylophone, (b) single notes on a piano, (c) a piano chord, (d) the clarinet played staccato as the cat and (e) the bassoon as grandfather in *Peter and the Wolf.*

Finally we should look closely at the wave form during the decay period to see whether it merely shrinks in amplitude or whether there are changes in form. 3.31 shows an expanded version of the trace of the harpsichord note of 2.5 c at three successive moments in time. The first is immediately after the transient has disappeared, the second is roughly in the middle of the decay time and the third is near the end. The changes in shape are obvious and the corresponding frequency analyses show clearly what is happening; the various overtones decay at different rates and so the quality is actually changing as the note fades away. The effect is well known to composers and all kinds of subtle uses are made of it in music. In particular it provides one clue to the solution of the problem that used to be a bone of contention between scientists and musicians – the amount of control the pianist has over the tone produced and to which we shall return in the next chapter.

READY FOR THE JOURNEY

It has taken three chapters to reach the real point of departure of the sound waves on their journey, and in the last section we have really been looking at the envelope or suitcase within which all the information later to be interpreted by our ears and brains is packed. Later we shall consider just how well the information is packed and what hazards can affect it in transit and during the unpacking stage. Our discussions have been somewhat theoretical, however, and it would seem to be a good moment to stop to consider the very practical problems and highly ingenious solutions that have cropped up during the long history of musical instruments.

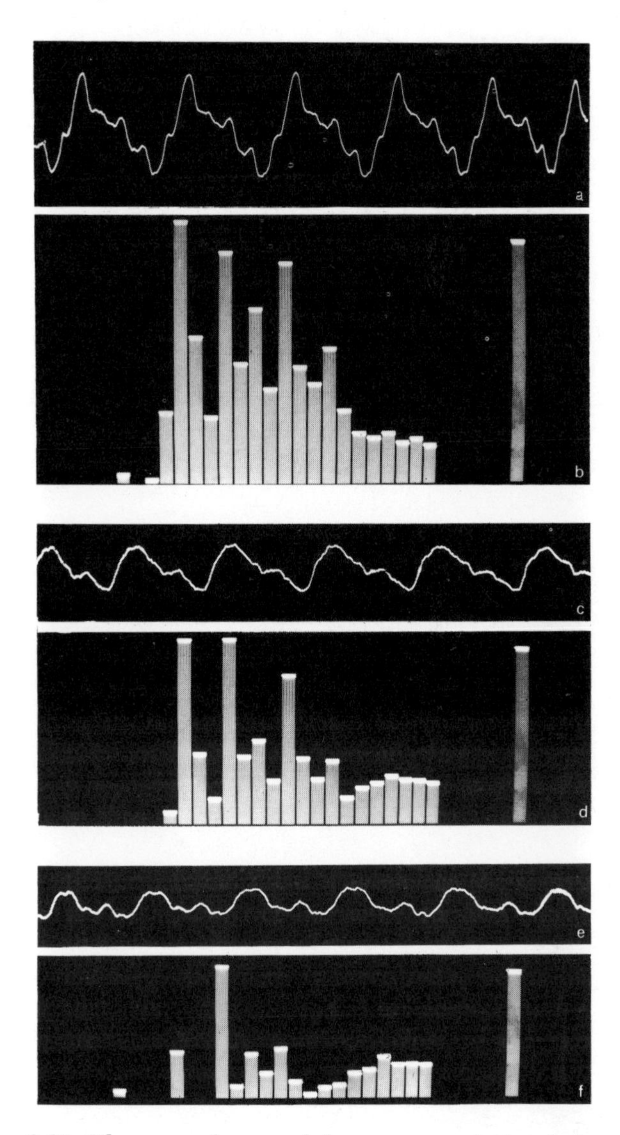

3.31 *The wave form and frequency analysis of the note C (262 Hz) produced on a small harpsichord in (a) and (b) immediately after the part of amplitude has been reached; in (c) and (d) about a quarter of a second later; and in (e) and (f) about 3/4 of a second later*

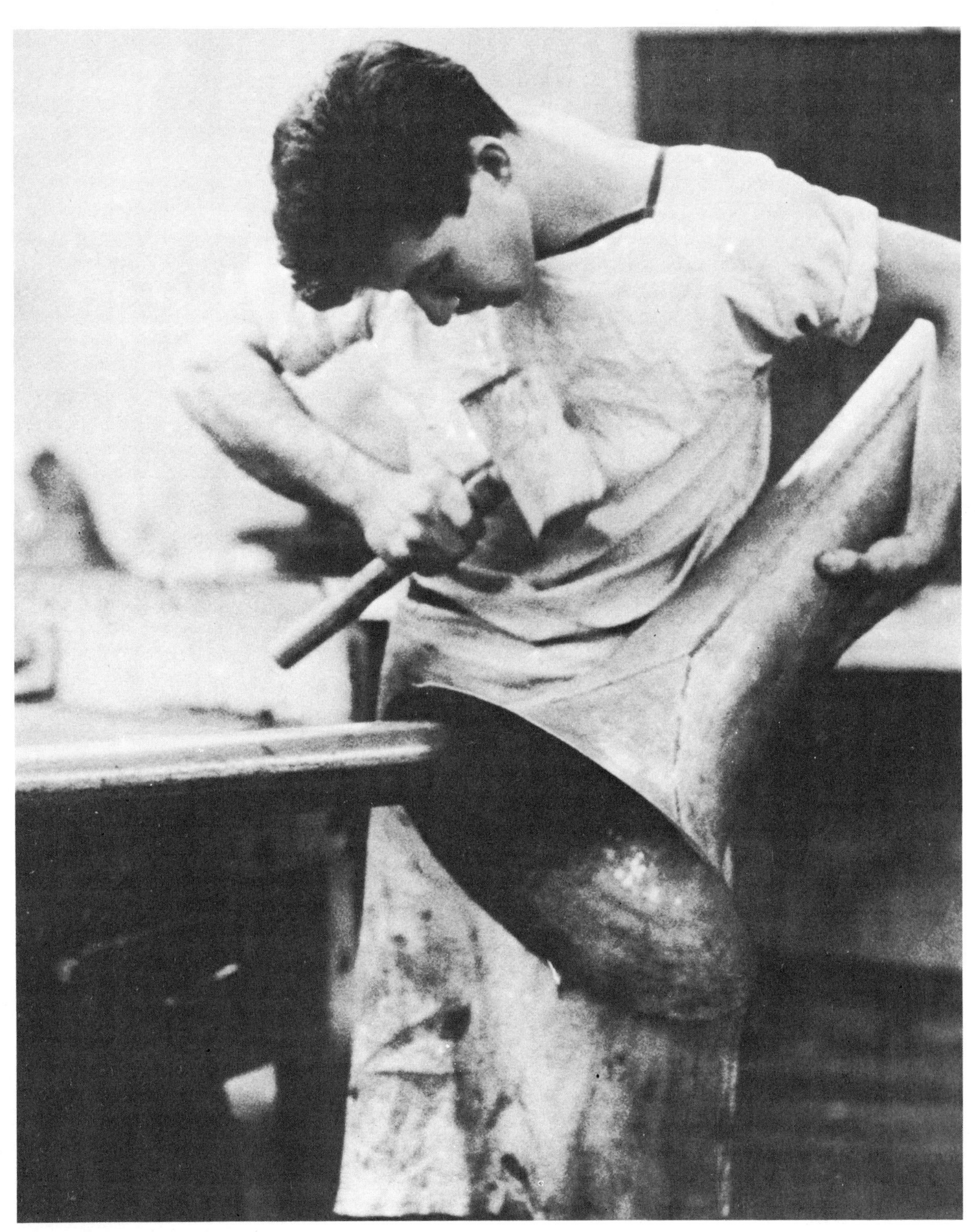

4.1 *Manufacturing the bell of a brass instrument*

CRAFTSMANSHIP AND TECHNOLOGY

INTRODUCTION

This chapter is already long, but could all too easily have become as long as the rest of the book put together! It takes us into the fascinating world in which the superb skill and ingenuity of craftsmen, who love their materials and tools almost as much as their finished instruments, interact with the artistry and creative imagination of composers and instrumentalists. A world, too, in which new techniques of scientific measurement and new applications of electronic technology are having an increasing impact and are enabling us to create totally new sounds. Studies that overlap many different areas of specialism are always attractive and this one is no exception. We must, however, resist temptation and restrict ourselves to a few selected topics chosen to illustrate some highlights of the amazingly fruitful partnership between Art, Science, Craftmanship and Technology. A great deal has been left out and I must apologise to any reader who feels that his pet subject has been undervalued, or even omitted altogether.

The blend of old and new manufacturing techniques will be illustrated by a brief look at the way in which brass instruments are made. The ingenuity of musical instrument makers will be illustrated by a short account of the mechanisation of woodwind instruments. The string family provides us with opportunities of observing the standards of precision and care demanded of instrument makers and will also give us a basis for considering ways in which traditional instruments are being redesigned and improved with the aid of scientific techniques. Keyboard instruments have been sources of controversy of one kind or another throughout their history, and new methods of analysis are beginning to shed light on some of their mysteries. The first successful application of electronics in a serious musical instrument was in the keyboard family and the pipe *v.* electronic organ controversy provides fascinating illustrations of the consequences of principles outlined in earlier chapters.

Moving then to the new sounds of music, the French musician-inventors of 'Les Structures Sonores' illustrate for us up-to-date methods of producing new instruments which are entirely mechanical and which again are based on many of the principles elaborated in earlier chapters. A few years ago large computers seemed likely to have a major part to play in the exploration of new sounds, but already they are being overtaken by powerful combinations of small computers and large electronic synthesisers and the chapter concludes with a brief account of the general principles on which the new systems operate.

These, then, are the topics to be touched on; let me emphasise again, however, that the intention is merely to indicate what can be done, to provide a glimpse of the tips of eight or nine fascinating icebergs, any one of which would merit a book to itself. We shall look first at wind instruments.

Conventionally we divide wind instruments into the 'brass' family and the 'woodwind' family, but the distinction is not made on the basis of the material of construction. Brass instruments are those which use the lips of the player as reeds and have a mouthpiece which is a more or less cup-shaped orifice across which the lips can be stretched. Most modern members of the family are made of metal, but various ancient members such as the serpent and medieval cornett are found made of wood, ivory and other materials. The woodwind, on the other hand, includes those instruments which have mechanical reeds, such as the clarinets and oboes, or are excited by edge tones, such as recorders and flutes. The material of construction is just as diverse; one finds modern flutes more commonly made of metal than of wood, and the saxophone, which is always made of metal, is often classified as a 'woodwind' because it uses a mechanical reed similar to that of the clarinet.

THE BRASS FAMILY

A year or two ago I had the opportunity of visiting the factory in which Messrs Boosey and Hawkes make all kinds of brass instruments ranging from the modern cornet to the BB♭ tuba and the Sousaphone. The mixture of craftsmanship and technology involved is fascinating. On the one hand the method of creating the flared horn still depends almost entirely on the skill of the craftsman beating out the shapes from sheet metal largely by hand; 4.1 shows such an expert at work. On the other hand hydraulic technology is involved in the semi-mass-production methods of producing the complex bends, crooks and joints that form the heart of the system of length-changing needed to permit brass instruments to play a full range of notes. You will recall that in chapter one we placed the brass instruments in family number three – those that change their pitch largely by changing the mode of vibration of the air in the pipes. Disregarding the extra complications that arise from the fact that the pipes are neither wholly cylindrical nor wholly conical as was mentioned in the last chapter, it is still clear that mode-changing is likely to leave quite large gaps in the musical scale – especially between the notes of the first few low-pitched modes (1.29 makes this point clear). Four techniques have been used at some time or other to overcome this problem. The first involves effectively shortening the tube length by using side holes as is done in the woodwind. The other three all involve lengthening the tube – a device which does not seem to have been tried with woodwinds except to make very minor changes in overall pitch for tuning purposes.

The medieval cornett was probably the earliest instrument of the brass family which could play chromatic scales as distinct from bugle and horn calls using solely the natural overtones of the tube. The origin is unknown, though pictures of instruments which clearly are cornetts survive from the thirteenth and fourteenth centuries. The oldest surviving specimens appear to date from about the sixteenth century. 4.2 a shows a seventeenth-century example.

The serpent is really just a large bass cornett which is bent into the characteristic shape to make it possible to reach the finger holes in relative comfort. It appears to date from about the end of the sixteenth century. 4.2 b shows an eighteenth-century version; later extra keys were fitted but the instrument was largely superseded by the bass horn or Russian bassoon which used a straight tube doubled on itself, as does the orchestral bassoon (see later section), and was therefore easier to make and hold than the serpent, but it retained the lip reed and so remains in the 'brass' family. All the instruments so far described are usually made of wood or horn – though examples of copper serpents exist. The first instrument commonly made of metal to which the finger-hole system was applied was

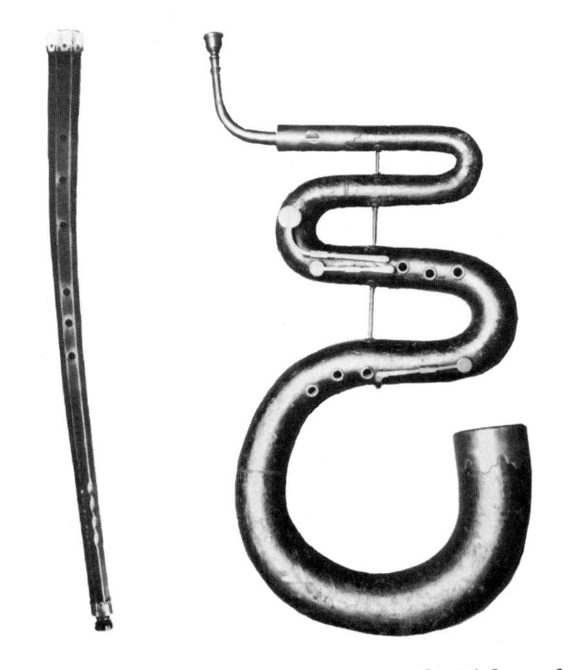

4.2 *(a) Seventeenth-century cornett. (b) Eighteenth-century serpent*

the keyed bugle and later a bass version known as the ophicleide. Of all this group only the ophicleide has ever made any significant contribution in large orchestras, and nineteenth-century composers such as Wagner and Mendelssohn have written parts for it – though they are nowadays played by tubas. The fundamental problem with this group of instruments is that it is extraordinarily difficult to find hole positions which are a sufficiently good compromise to enable notes to be played in tune while retaining some constancy of tone over a wide range. The positions can be set right for one particular mode – say the second harmonic – but they will then be incorrectly placed for higher harmonics and the combination of complex fingering and 'lipping' becomes unmanageable.

The second approach was to provide alternative lengths of tubing or 'crooks' which could be plugged in at will to enable the whole pitch of a bugle or horn to be changed, but this clearly cannot be done rapidly enough for playing tunes!

The third approach – used in the now familiar trombone – is to use a sliding crook which enables the performer to lengthen the tube continuously over a fairly wide range. The big advantage of the trombone is that a skilled player can adjust the length to precisely the right amount for any harmonic without any need to compromise. On the other hand, to make a long, air-tight slide which will, nevertheless, move very freely is quite a challenge to the instrument maker.

The fourth solution, used on all modern orchestral instruments of the brass family except the trombone, is to fit three or four crooks which can be brought into use as desired by depressing pistons which control valves; they effectively add extra specific lengths to the total length of pipe already there. On three-valve instruments the extensions of length provided correspond respectively to a tone (valve 1), a semitone (valve 2) and one and a half tones (valve 3). Thus the greatest gap in the harmonic series above the second harmonic – the interval of a fifth between the second and third – can be filled. For example, if the second harmonic is A the third will be E and the semitones between can be produced by exciting the third harmonic and depressing the valves as follows:

Note required	*Valves depressed*	*Total lowering of pitch*
E	—	0
D♯	2	$\frac{1}{2}$ tone
D	1	1 tone
C♯	3	$1\frac{1}{2}$ tones
C	2 + 3	2 tones
B	1 + 3	$2\frac{1}{2}$ tones
A♯	1 + 2 + 3	3 tones
A	—	but change to second harmonic

The problems in practice, however, are quite considerable because clearly, from our discussions of the relationships between pitch and frequency in chapter one, we can see that to create a change of a given interval – say a whole tone – needs a given *percentage* change in the length of the tube. A major tone corresponds to the frequency ratio 9:8 (see appendix), so to lower the pitch by a tone needs an extension equal to one-eighth of the existing length, i.e. of $12\frac{1}{2}$ per cent. If the tube is, say, 100 cm long this corresponds to an extra $12\frac{1}{2}$ cm; but if the pitch has already been lowered (valve 3) by $1\frac{1}{2}$ tones and therefore the total tube length is already about 119 cm, the additional $12\frac{1}{2}$ per cent is nearer 15 cm; thus if we only lengthen the tube by $12\frac{1}{2}$ cm the resulting note will be about one-third of a semitone sharp. A player can 'pull' the note flat or sharp by altering the air pressure and lip tension to some extent, but a third of a semitone is a large amount and the tone would certainly suffer. With the higher-pitched instruments, like the trumpet, it is not impossible for a skilled player to dominate the instrument – but with a double bass tuba the volume of air involved is so large that the odds are very much on the side of the instrument! This is where the ingenuity of the instrument maker comes in. He can make the crook for the third valve a compromise length so that it is wrong for all situations, but the errors are evened out, or he can use a 'compensating' system which is a most ingenious device which, when valve 3 is depressed, adds a little bit extra to the crooks of both valves 1 and 2 in order to get the percentage more nearly correct.

It will have already become obvious that the gap between the fundamental – or pedal note as it is usually called – and the second harmonic cannot easily be covered with a three-valve instrument. Many of the larger brass instruments are now fitted with a fourth valve which lowers the pitch by $2\frac{1}{2}$ semitones. Hence if all the valves are depressed a total lowering of $5\frac{1}{2}$ semitones is possible and this corresponds to the eleven chromatic semitones between the second harmonic and the fundamental. Our comments about changing percentages, however, become even more important with four-valve instruments and fully compensating systems become essential. 4.3 a shows diagrammatically the ingenious way in which a three-valve compensating system works and 4.3 b is a close-up of the valves, crooks and compensating loops of the same system in a practical instrument. I am sure that we have already delved deeply enough to indicate the complexity of problems of principle that arise – and we have hardly begun! Now consider the practical problems of manufacture of this maze of pipe-work. Consider also (in the light, for example, of 3.23 and the corresponding discussion in the last chapter) the problem of creating all these joints and bends without sudden changes in the diameter of the bore and you will begin to see just a little of the task which the instrument maker faces.

Of the many fascinating techniques involved I shall just choose one – the hydraulic technique for producing tubes with very sharp bends. Try bending a piece of fairly thin-walled brass tubing and see what happens. The result will almost certainly be as shown in 4.4 a. In the Boosey and Hawkes factory, however, the technique used is deliberately to 'dimple' the tube first (4.4 b), then to bend it into the required shape (4.4 c) and finally to expand it back to the required shape by placing it in a split mould (4.4 d) and forcing oil into it under enormous pressure. This is an over-simplification and various annealing and other processes must be gone through – but the result is the beautifully smooth pipe-work already seen in 4.3 b.

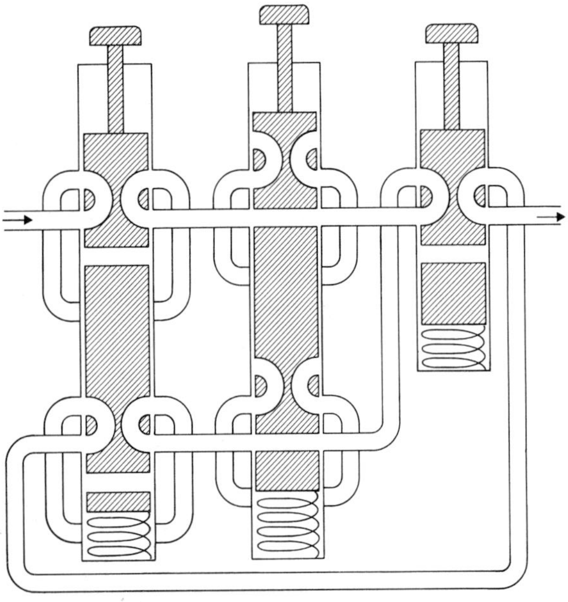

4.3 *(a) Diagrammatic picture of a three-valve compensation system for a brass instrument. (b) Close-up of an actual compensation system in a tuba*

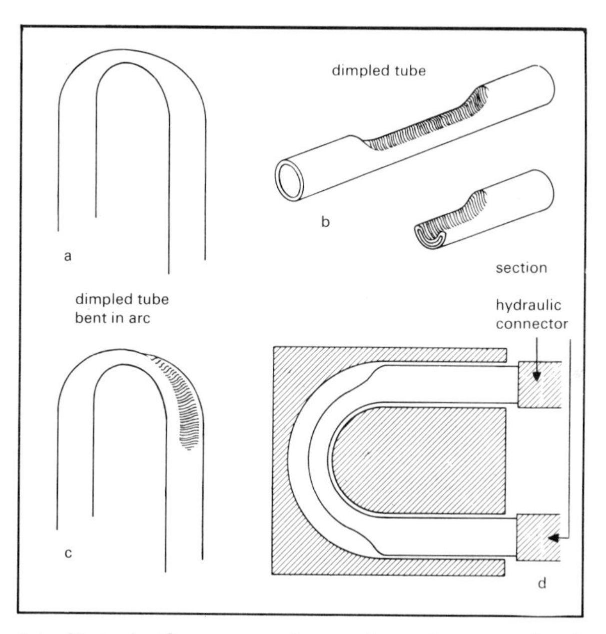

4.4 *Steps in the process of manufacturing acute bends for brass instruments*

The mouthpieces of brass instruments are real works of art and most players have their own favourite design. The scientific function is to provide a frame across which the lips are stretched, and to provide an efficient transfer of the vibrations set up in the air by the lips into the end of the instrument itself. The mouthpiece is not a uniform tube and behaves rather like a Helmholtz resonator; because of the increase in diameter of the cup it adds a greater effective length than its actual measured length and the 'end effect' is different at different frequencies. In practice the higher modes are flattened in pitch more than the lower ones, so it plays an important part in modifying the pattern of overtones both in the 'vibration recipe' of any given note and the actual notes produced. Although some studies of the way in which a player's lips behave, and of systematic changes in shape, have been done the design of mouthpieces is still largely empirical and the best mouthpieces are probably produced by designers who themselves are accomplished performers.

We discussed briefly in the last chapter the effect of the bell in radiating the sound out of the tube, and in determining the cut-off frequency below which waves are reflected back to the reed, so providing both the necessary loudness and a characteristic formant that helps to establish the tone quality of a particular instrument. Just as the mouthpiece affects the frequency ratios of the overtones, so does the bell – so incidentally do any slight variations in bore. It is even possible, as we saw in the last chapter, by deliberately producing a constriction or expansion at a particular point to change the pitch of certain modes or resonances and not others. What a complex system and what an enormous number of variables there are to play with! No wonder the simple treatment of harmonics of tubes open at one or both ends does not seem to match up with the practical behaviour of brass instruments.

THE WOODWIND FAMILY

In some ways the woodwinds – at least in their modern form – behave in a rather more predictable way. This is largely because they tend to be either almost uniformly cylindrical (as the clarinet or flute) or uniformly conical (as the oboe or saxophone). Our particular interest here, however, is to look at the ingenuity of the systems of keys that have been developed in order to increase the range and flexibility of the notes produced. First let us consider the simplest form of pitch-changing mechanism. The earliest woodwinds that still survive almost unchanged are the recorders and we shall take the treble or descant recorder as an example of the simplest mechanism. Flutes and recorders are cylindrical pipes effectively open at both ends and hence – as we saw in earlier chapters – will overblow into the second mode which is the second harmonic, i.e. one octave above the fundamental. To fill in this gap of one octave to produce a diatonic scale we need seven successive steps of shortening. This is provided by six holes on the front face of the instrument and one on the back. 4.5 a shows the system diagrammatically and 4.5 b shows how the holes are covered. This system is still at the core of most woodwinds, even of the most sophisticated keyed instruments in use today. Semitones can be provided either by half

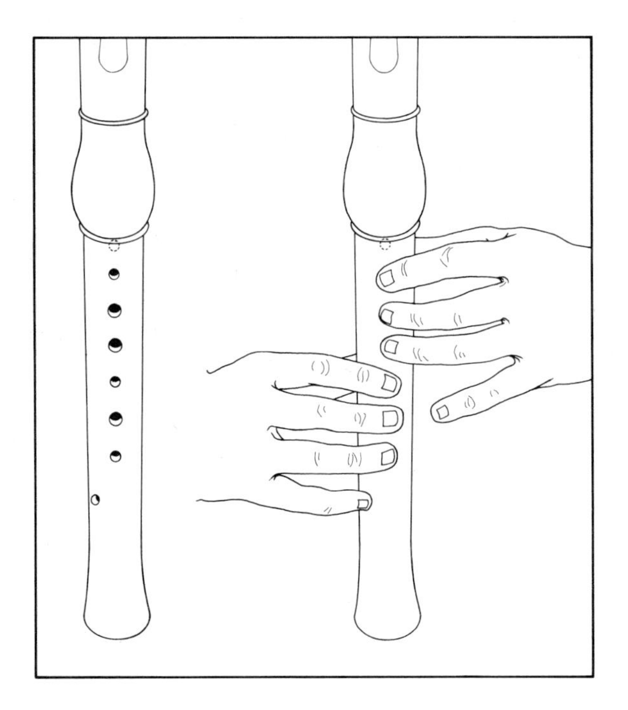

4.5 *Hole positions on a recorder*

covering holes or by closing one or two holes *lower* than the one defining the basic vibrating length of the tube. The second octave and even higher notes can be produced reasonably in tune by overblowing. But complex systems of 'cross fingering' are often introduced for higher notes; essentially by opening and closing holes in a certain pattern, which relates to the positions of nodes and anti-nodes, the pipe can be encouraged to stay in a particular higher mode of vibration.

Keys are added for four main reasons in traditional instruments:

(a) To enable holes to be covered or uncovered that cannot otherwise be reached.

(b) To enable more holes to be controlled than is possible with the fixed number of fingers available to human performers.

(c) To enable larger holes to be covered than is possible with normal fingers.

(d) To permit the performance of very rapid passages or trills which would otherwise involve such complex changes of fingering that fluent execution would be almost impossible.

There is a fifth reason which probably would not be accepted by orthodox musicians, though one example will be given; that is, to make it possible to play without learning complex and apparently illogical fingering systems. The solution is to design an electronic or electromagnetic key system which really transforms the instrument into a 'one key – one note' system like a piano. It would be out of place to give a complete history of the development of key systems, but the following outline should be adequate to indicate the principles.

There are four basic types of key used on instruments of the woodwind family, though they may appear in complex combinations:

(a) Closed keys (4.6 a), in which pressure on the plate opens the hole.

(b) Open keys (4.6 b), in which pressure on the plate closes the hole.

(c) Hinged keys (4.6 c), which merely provide for the closing of a larger hole than would other-

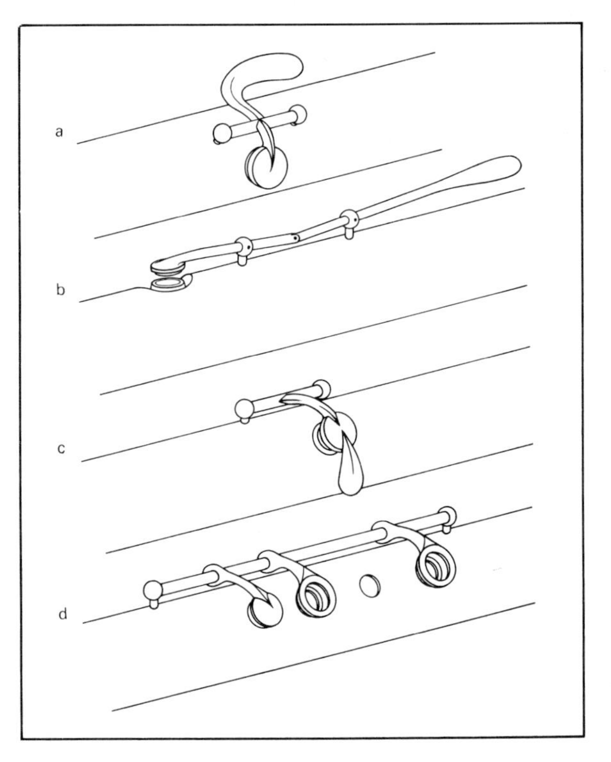

4.6 a–d *Types of keys on woodwind instruments: (a) closed key; (b) open key; (c) hinged key; (d) ring key*

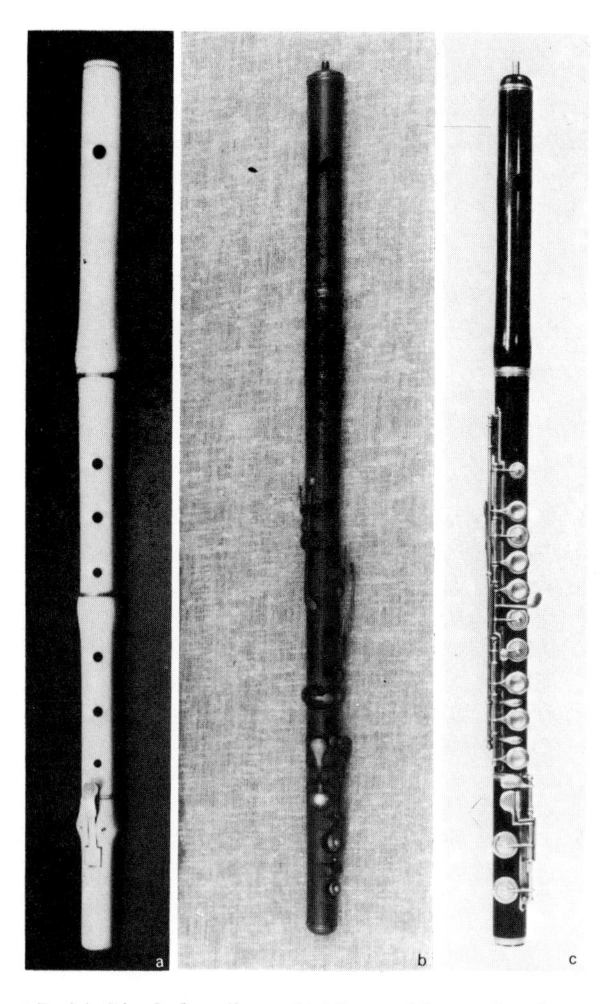

4.7 *(a) Single-key flute. (b) Flute with two foot keys. (c) Boehm flute*

wise be possible. Sometimes these keys can be pierced, so that by moving the finger to one side the main part of the hole is blocked but a smaller hole remains with consequent intermediate pitch change.

(d) Ring keys (4.6 d), which permit simultaneous control of more than one note by the same finger. These keys form the basis of the Boehm mechanism which revolutionised the woodwinds.

Closed keys were probably the first to be used, often to provide a semitone between the note produced by a flute or recorder when all holes are closed and that when the lowest finger hole is uncovered; treble and descant recorders manage this with an ordinary hole for the little finger, but on those of lower pitch and on the flute the hole turns out to be too far away for the little finger of the right hand to reach. 4.7 a shows an eighteenth-century flute with such a key.

Open keys were first used to extend the range downwards. Thus with a flute whose normal lowest note is D it might occasionally be useful to go down to C♯ or C, and this can be done by making the tube longer in the first place and bringing it back to the same effective length on all the notes in the normal range. They have overlapping plates and one or both can be pressed by the little finger when notes below the normal D are required. 4.7 b shows such a flute made in the late eighteenth century; it also has three closed keys higher up to give sharps or flats more positively than by half covering holes.

The real transformation came from Theobald Boehm in the middle of the nineteenth century. It is especially interesting in view of the general theme of this chapter to note Boehm's background. He was a jeweller by trade, used to the kind of fine detailed metal work that now characterises instruments based on his designs. He was also a flute player of considerable accomplishment who occupied principal positions in professional orchestras. Finally he had a scientific turn of mind and was not disposed to accept all the myths and traditions of existing instrument makers if revolutionary changes could produce positive improvements in tone or execution. There seem to have been two main problems with which he concerned himself. The first was the rather dull and poor quality associated with so-called fork-fingering of semitones – that is, lowering the pitch of a given note by covering the *next but one hole* below. The second was that flute makers in the past used relatively small holes, placed them in the most comfortable position for the player and then changed their relative sizes to bring the notes into tune. You will recall that, in the last chapter, we stressed that the sound comes out of the finger holes rather than from the open end, so they have a dual purpose. Boehm felt that they should be bigger and should be placed in such a way as to get the best and most uniform tone consistent with tuning, re-

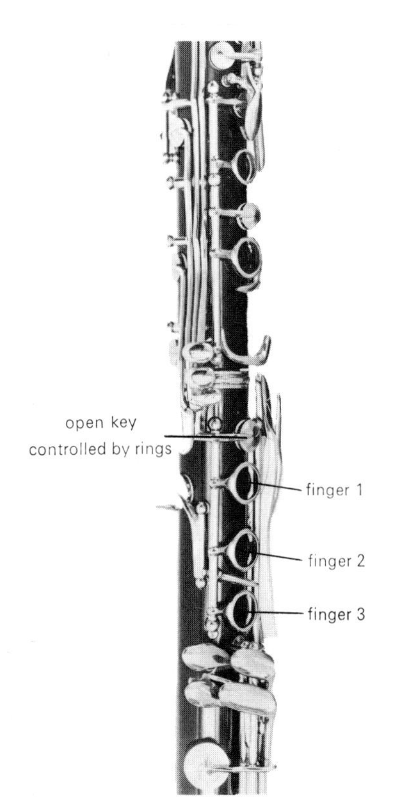

4.8 *Ring keys on a clarinet*

gardless of comfort. As a result of empirical experiments it became clear that without keys such a flute would be unplayable – partly because of the large size of the holes and partly because of their position. Hence the Boehm mechanism and the Boehm system of fingering. Again we shall not go into great detail – 4.7 c shows a modern flute using the Boehm mechanism. The basic idea is that the fingers should not have to move away from their positions controlling the main holes and that elaborate linkages and a combination of all four types of keys mentioned above should be used to permit the fingers full control with a minimum of movement. Typical of the ingenuity that went into the system is the use of ring keys in effect to provide the player with an extra finger.

Suppose there are four finger holes to be operated with only three fingers (see 4.8). The highest hole is fitted with an open key that can be closed by depressing any one of the three ring keys which are placed over the remaining three holes. Now if all three fingers are down, all four holes are closed. If finger 3 or fingers 2 and 3 are raised the corresponding holes are opened, but finger 1 keeps the highest hole closed as well as its own. If finger 1 is raised and finger 3 depressed, finger 3 keeps the highest hole closed via the ring key and this is far enough away to avoid influencing the pitch or tone. Finally if all three fingers are raised, all four holes are uncovered. Hence four holes with three fingers without any sideways movement at all! This kind of principle extended and modified in different ways forms the basis of the mechanism of most modern woodwinds. It should be clear now how much easier it is for the player to play rapid passages when only up-and-down movements of the fingers are needed and sideways motion is cut to the minimum. There are many more refinements possible. For example, Boehm himself devised a new type of key which acted as a combined ring and hinged key. With this one can use three fingers for four holes as described above, but also larger holes can be used than could be satisfactorily closed with the fingers alone. The flute of 4.7 c is fitted with this system. It is almost out of the question to perform a trill with the thumb hole or thumb key – since the thumb is needed to hold the instrument steady – hence the addition of alternative 'trill keys' for certain notes. On the clarinet – which as we saw in the last chapter has a cylindrical bore with the reed acting as a closed end – the first overtone is the *third* harmonic (a twelfth above the fundamental) and hence a more complicated fingering system is needed to cover the gap between the octave and the twelfth. 4.9 shows parts of a modern oboe and a clarinet on which the various types of keys can be seen (including pierced keys on the oboe). On the clarinet the variation in hole size can clearly be seen.

Two points should be mentioned before we drag ourselves away from this absorbing topic; the tenor member of the oboe family – the bassoon – provides the basis for both. As instruments become larger the finger holes necessarily become further apart and three adjacent holes cannot be placed close enough to be covered by three fingers of the same hand and still play in tune. The bassoon maker solved this by adding a very thick wall (called the wing) at the appropriate place and drilling the holes at an angle so that inside the bore they are the correct distance apart to give the right notes and outside they are close enough together for the fingers of one hand. 4.10 a shows an external view and 4.10 b is an X-ray photograph of the wing showing clearly the oblique drilling. Needless to say this arrangement plays its part in modifying the tone of the bassoon. Indeed it turns out that the side holes in all woodwinds have a surprising effect even when closed. It seems that the behaviour which led to a certain recipe inside the pipe for an absolutely smooth bore is distinctly different from that of the bore with regular 'bumps' in it which results from closing the side holes with the fingers or keys applied only to the outside of the bore. The cut-off frequency

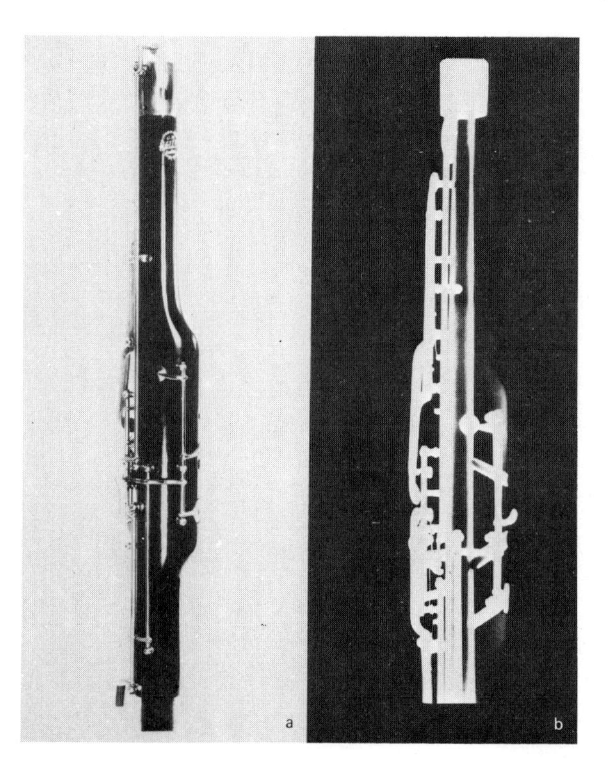

4.10 *(a) 'Wing' section of a bassoon. (b) X-ray photograph showing oblique drilling of finger holes*

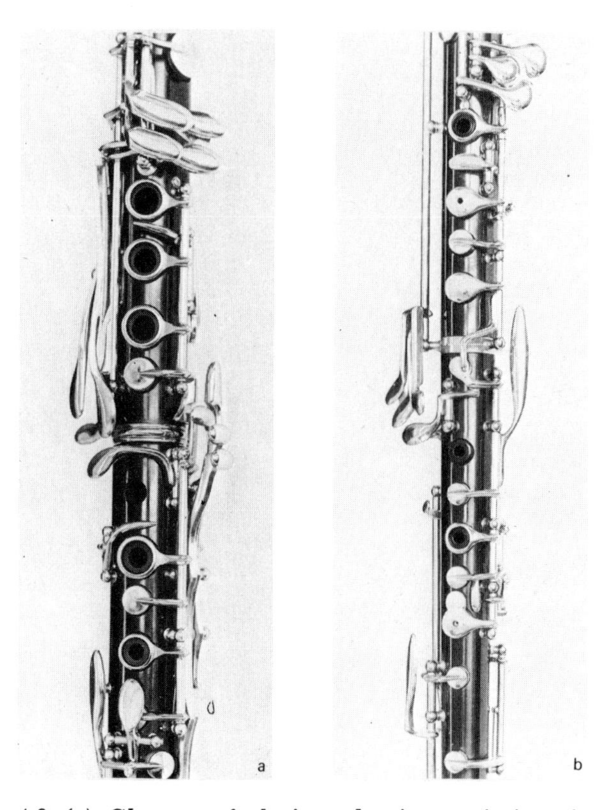

4.9 *(a) Close-up of clarinet showing variations in hole diameters. (b) Close-up of oboe including pierced keys*

is modified by these bumps, though not to quite the same extent as by the regular succession of open holes.

The second point is that it will have become abundantly clear that fingering patterns on modern woodwind instruments have moved a long way from the simple raising of the fingers in order used, say, for the lowest octave on a recorder. Can this complication be overcome?

Dr Giles Brindley has developed what he calls a 'logical bassoon' in which the fingering pattern for each note is set up automatically by depressing one key. Such a device is bound to be viewed with some misgivings by performers but, on the other hand, it may present the possibility of obtaining enjoyment out of playing a difficult instrument to some who would not have the time, patience or manual dexterity to master the conventional method. So much then for our glimpse of the complexities and problems involved in making and playing some of the woodwinds.

TRADITIONAL STRINGS

From some points of view it is probably true to say that the making of violins has had greater attention from historians, musicians and scientists than any other family of musical instruments. Stringed instruments in general are visually very attractive; they are made from wood which may have a beautiful grain; older examples have a varnish which brings out the depth of the colour and markings and gives the instruments an almost translucent richness; even the external shape has a beauty of form and often elaborate decorations are added. 4.11 shows three superb examples of the art – a sixteenth-century Venetian guitar, a viola d'amore from the eighteenth-century and a Stradivarius violin. But most remarkable of all is the fact that the slow painstaking trial-and-error efforts of violin makers led to the development of shapes, sizes and materials which have remained virtually unchanged for at least 300 years. Great strides have been made in scientific studies of violins during the last twenty years or so – but, though it is true that it is now possible to make instruments with required tonal characteristics which are very good indeed, the very best of the instruments produced in the Stradivari–Amati–Guarneri golden age have never since been equalled.

In earlier chapters we outlined what the body of the violin is really required to do. It must take the vibrations of any one of the four strings at any of the frequencies within the wide range

4.11 *(a) Venetian guitar from the sixteenth century. (b) Viola d'amore from the eighteenth century. (c) Stradivarius violin*

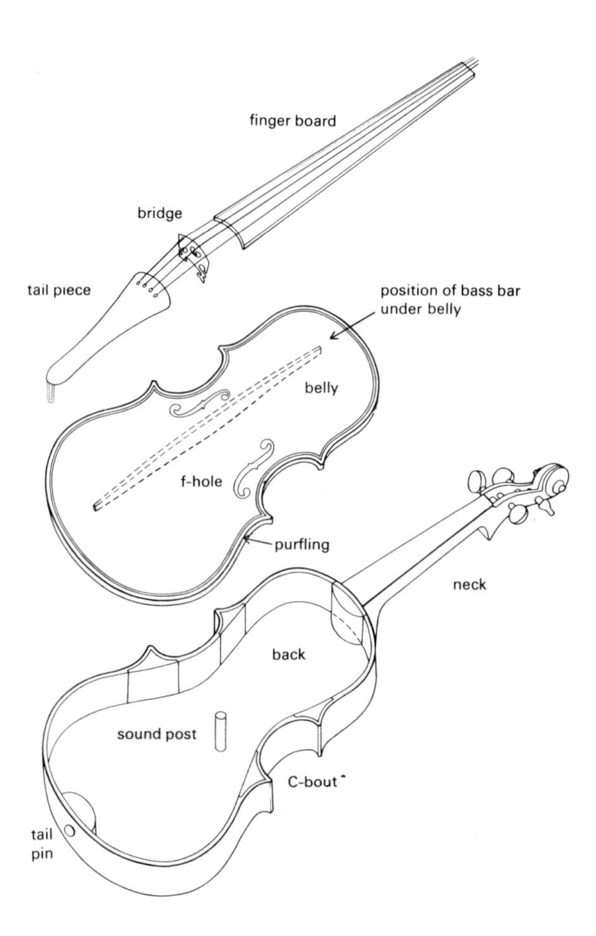

4.12 *Parts of a violin*

of the instrument and amplify them, and all their relevant overtones, in such a way as to give a uniform blend of tone. It does this by means of a box of highly complex shape, with walls that are very non-uniform in cross-section, and which has odd asymmetries (such as the sound post connecting the back and belly approximately under one foot of the bridge, and the bass bar, a strip of wood glued to the belly more or less parallel with and below the lowest string).

4.12 shows diagrammatically the components of a violin or 'cello with the names of the principal parts. Before going any further with the scientific aspects of the design it would be good to pay tribute to the craftsmanship that has gone into this marvellous family by showing some of the stages in the making of an instrument by a modern craftsman. The instrument is actually a 'cello and the craftsman is Alec McCurdy who works near Newbury, Berks.

4.13 a shows him beginning to mark out the back plate. The block of wood, which has been carefully selected from a stock of well-seasoned planks, is formed from two pieces glued together along the centre line to achieve bilateral symmetry in grain – which has acoustic value as well as visual appeal. The half template is used to ensure symmetry in shape. 4.13 b shows part of the shaping process by which the beautiful and critically important convex shape is achieved. The plate for the belly is of a different kind of wood with a straight grain running along the length of the instrument and, although it is made of a single piece of wood, is produced in exactly the same way by tedious and exacting shaping with hand planes. 4.13 c shows a belly plate nearing completion – the straight grain, the *f*-holes and the convex shape can all be seen; the gauge is being used to measure the thickness at various places because the internal and external curves are not the same. 4.13 d shows a completed belly being fitted with the bass bar.

The ribs or walls of the instrument are made from thin strips of wood which are bent to shape using a hot metal roller. In 4.13 e one of the two 'C'-bouts is being bent; other completed ones can be seen standing on the bench. 4.13 f shows a stage in the assembly and, in particular, we can see the block of wood which strengthens the bottom end of the body and provides both the anchoring point for the tailpiece to which the strips are attached and the socket for the telescopic leg on which the 'cello is stood when played. 4.13 g shows the final stages in glueing and clamping the body and in 4.13 h the groove for the purfling – an inset sandwich of three thin strips of wood of which more later – is being cut. Also in this photograph can be seen some of the very tiny planes which are used in the process of shaping the complex curves of back and belly. Then the neck, the finger board, the tailpiece, the bridge, pegs, etc., all made from different types of wood chosen to perform a particular task, must be completed and assembled – and finally comes the varnishing, which may take many months as each coat must dry for

a

b

c

d

e

f

g

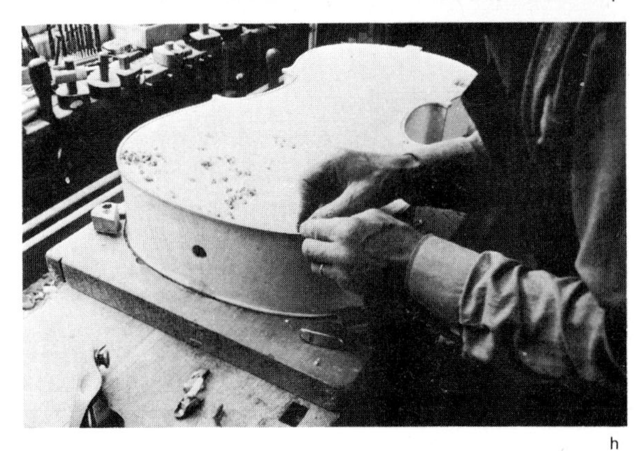

h

4.13 a–h *Alec McCurdy making a 'cello*

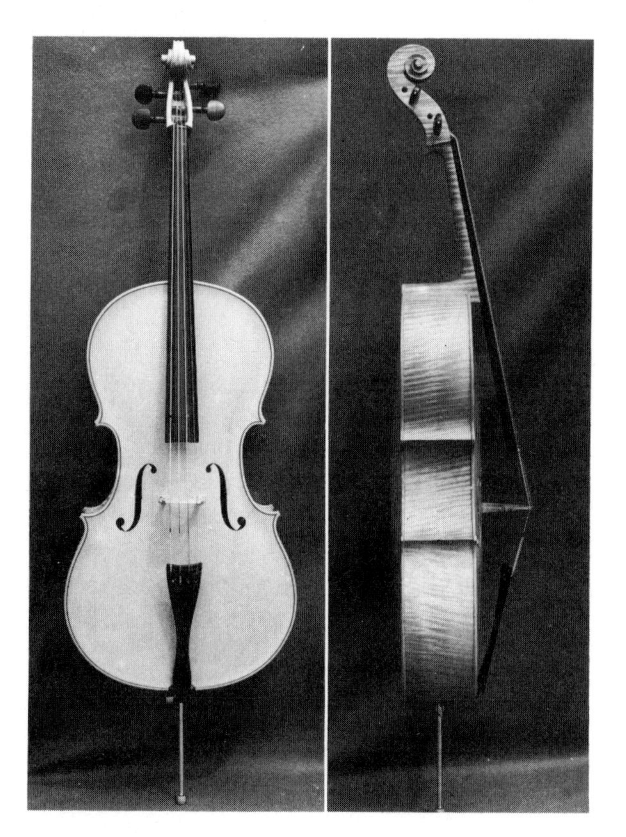

4.14 *Complete 'cello*

several weeks before the next is applied. 4.14 shows a completed instrument.

A NEW LOOK AT THE FIDDLE FAMILY

Wherein lies the secret of the tone? The varnish has often been held to hold the key, but the present-day view is that, while poor varnish can ruin the tone of an otherwise excellent instrument, the best available varnish made to some secret formula handed down the generations will do nothing to improve the tone of an inferior instrument. Most makers would agree that the tone of an unvarnished instrument can be better than the tone when varnished – but the varnish is essential to protect and preserve the wood and, in particular, to prevent its properties changing rapidly as the humidity of the air changes. A good varnish should be light in weight, not too thick, not so soft that it will damp down vibrations and not so hard that it will make the instrument too rigid.

Most makers have their own favourite methods of testing and tuning the various components and there are legends of craftsmen wandering through the forest tapping trees to select wood with the right properties. In recent years the Catgut Acoustical Society of America has conducted and encouraged research by an ever-increasing group of musicians, scientists and fiddle-makers, into ways in which modern science can help. Their researches have not only led to some interesting and important new principles but have also enabled the society, with the dynamic leadership of its secretary, Mrs Carleen Hutchins, to design and make a totally new family of fiddles. The work was begun by the late F. A. Saunders, who was a professor of physics at Harvard.

Probably the central feature of all the work – which can only be described very briefly here – is the development of methods of surveying the resonant frequencies of the separate components and of the assembled instruments and finding relationships between them so that some of the trial and error can be minimised. One of the major problems, of course, is in deciding what sort of tone is desirable, and this highly subjective choice is bound to vary from player to player and from critic to critic. Clearly, too, the quality needed for chamber music in a smallish room is quite different from that needed for brilliant execution of a concerto in a large concert hall and with a large orchestra.

Saunders introduced a very useful semi-scientific method of assessing violins which involves playing a succession of about a dozen notes on each of the four strings in turn. Each note is played as loud as possible and the measured loudness is then plotted on a graph. Loudness curves are different for every violin. A major problem in trying to be scientific is that a really first-class player can make a poor violin sound quite acceptable and it is quite difficult to be sure that each note is really being played as loud as possible. Poor violins in general have much more uneven loudness curves, and top-class violinists have a kind of built-in compensating system which enables them to use their bowing skill to even out the variations. A beginner, on the other hand, cannot cope with wide variations. I suppose this could be taken as an argument in favour of letting beginners learn on a Strad rather than the usual semi-mass-produced school-type instrument. I suspect that there are counter-arguments on non-musical grounds, however!

Hundreds of tests on thirty or forty different instruments were made and some general principles have emerged. The best of the old violins (e.g. among others the Guarnerius of Jascha Heifetz) have two of their principal resonances – the so-called wood or main body resonance and the air or cavity resonance – not far from the frequencies of the open A string (second string – 440 Hz) and the open D string (third string – 294 Hz) respectively. The main body resonance is critically dependent on the precise thickness of the wood at all parts, on the bass bar, and many other parameters. It has already been found that the removal of almost infinitesimal amounts of wood from the bass bar, or some other critical point, can make very significant changes in the resonances and well-made instruments with rather poor tone have already been taken apart and rebuilt after suitable shaving of the bass bar or other parts to give improved tone. The purfling (the sandwich of thin strips of wood glued into the groove round the outer rim of the back and belly mentioned earlier) – originally provided to prevent the spread of cracks at the edge of the instrument – can play quite a significant part in determining resonances; the effect is greatest if the purfling is a bit loose so that effectively the back or belly has a very thin edge. The cavity resonance depends on the depth of the ribs, the extent of the arching of the back and belly, the size of the *f*-holes and again many other parameters.

The problem is to predict what the resonances of the completed body will be without having to wait until it is assembled and glued up. Again surveys of a wide range of instruments show that it is possible to predict in a more or less empirical way what the resonances will be if one measures the wood resonances of the back and belly separately. There is no theory as yet – and the likelihood of producing a really complete mathematical model is small because the resonance of the assembled instrument depends on things like the nature of the glue, the depth of the purfling, the mass of the bridge, the tension of the strings and many other interdependent factors. For a normal-size violin, however, it turns out that the back plate usually has main resonances on its own that are between a semitone and a tone higher than those for the belly on its own. If they are further apart the resulting quality is often poor though recent work suggests that this is not always the case. On most good instruments the mean of these is usually about three and a half tones below the main wood resonance when the fiddle is complete.

This indicates that the sequences of resonances for each are more or less interlaced. Indeed there is a close parallel between the cooperative phenomenon discussed in connection with the woodwinds in the last chapter and the parts played by the various resonances in the string family. One prominent resonance of the plates is usually called the 'tap tone', since it can be evoked literally by tapping them. This indeed is the method used by many violin makers. There are problems, however, since the way in which it is held and the point of tapping make significant differences to the specific resonance heard. Indeed the upper and lower parts of either plate have tones which are often distinctly different. The 'C'-bouts and the *f*-holes of the belly seem to divide it into two fairly distinct but coupled systems.

The discussion so far is certainly not likely to enable you to make your own fiddles. I hope, however, that something of the blend of art and science that is involved will have become apparent. We are now going to look briefly at two or three of the very latest scientific ideas that are being used to help the study of the string family.

One problem already discussed is that of ensuring symmetry and attaining the required curvatures of the surfaces of the plate. These are not simple cylinders or spheres but quite complex figures whose curvature in different

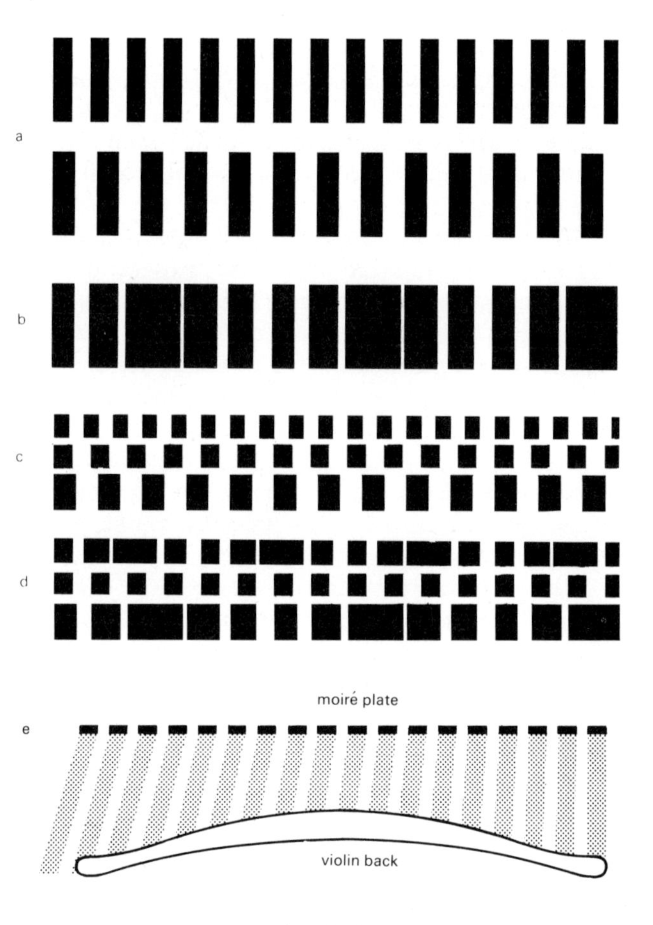

4.15 *Principle of moiré techniques*

directions varies from point to point. Among the main devices that have been proposed for assessing the progress in plate shaping, one of the most useful and elegant is that using moiré fringes. The principle is well known nowadays: 4.15 a shows two sets of black and white bands, the lower of which are slightly wider than the upper. 4.15 b shows what happens if we now place them on top of each other; at some places they coincide and at others they do not and hence dark bands appear. In this case the lower set of bands is uniformly longer than the upper and hence uniformly spaced bands appear. 4.15 c carries the idea one stage further. Here the upper set is the same as for 4.15 a, but the lower set changes in width from top to bottom and the resulting pattern 4.15 d gives bands which are related in a predictable way to these variations; they provide a kind of contour line. 4.15 e shows how this principle can be applied to violin plates; a point source of light illuminates a plate on which is printed a fine pattern of regularly spaced parallel lines on glass and it casts shadows on to the plate. When the plate is in contact with the glass the shadows of the lines will be exactly the same size as the lines, but everywhere else the shadow will be magnified more or less depending on the distance between the glass and the plate. When the observer looks through the plate he sees moiré patterns between the lines on the glass and their shadows; the result is a contour pattern of the plate. 4.16 shows two such patterns – one of which reveals regions in which the shape is not at all satisfactory and, in this example, it would present quite a problem as too much wood has already been removed at the top left and bottom right. (Needless to say this is a deliberate exaggeration using scrap plates to make the effect very clear.)

The second technique is to excite the plate by means of a vibration unit – rather in the way that we excited the Chladni plate in chapter three – and to measure the amount of sound radiated from the plate by means of a microphone placed a fixed distance away. The driver is fed from an electronic source of pure tones which can be varied continuously in frequency from

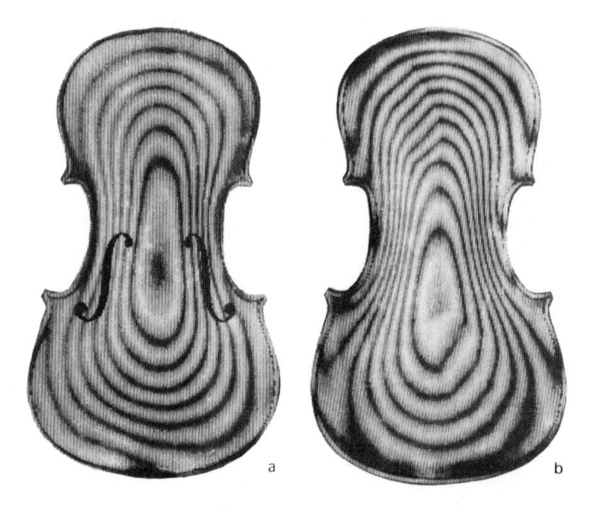

4.16 a and b *Moiré patterns of: (a) moderately symmetrical belly plate; (b) exaggeratedly bad back plate*

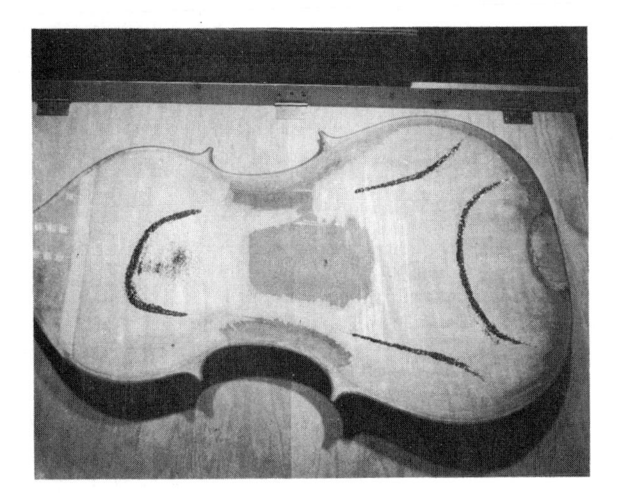

4.17 *Double-bass back plate excited by a 109-Hz tone from a loudspeaker underneath showing vibration patterns*

perhaps 30 Hz or so up to 20,000 Hz. The increases in loudness at specific frequencies denote resonances. It is possible to drive the plates vigorously enough to produce sand patterns showing the modes with a very high-power amplifier and a large loudspeaker unit placed below the plate. 4.17 shows the back of a double-bass revealing such a sand pattern. It should be remembered, of course, that such vigorous driving could lead to very different conditions from those experienced during playing.

The third technique to be mentioned briefly uses the same idea of driving the plate, but enables one both to detect resonances and to see the mode patterns when the amplitude of the driving signal is very small indeed and quite comparable with that likely to be experienced in playing. It uses the relatively new optical technique of holography.

The principle is demonstrated in 4.18. It is important first of all to understand that ordinary light waves are electromagnetic just as are radio waves, but whereas radio waves have frequencies measured in megaHertz, light frequencies are measured in hundreds of millions of megaHertz. Radio transmitters emit continuous trains of waves, but light, on the other hand, normally consists of very large numbers of short bursts of energy. In recent years the development of lasers has opened up all kinds of new possibilities, and for our purposes here it will suffice to say that the laser produces a beam of light that is not only very much more intense than any other, but also emits more or less continuous wave trains. It could be thought of crudely as a radio transmitter that happens to work in the frequency range of light.

If a beam of such light falls on an object, the light is scattered (4.18 a) and we could form an image as with ordinary light by putting a lens at L in the figure. If instead we place a photographic plate at L and reflect a part of the signal beam on the plate also (4.18 b) a remarkable effect occurs that would not occur with light from an ordinary source. Each tiny point on the object scatters light

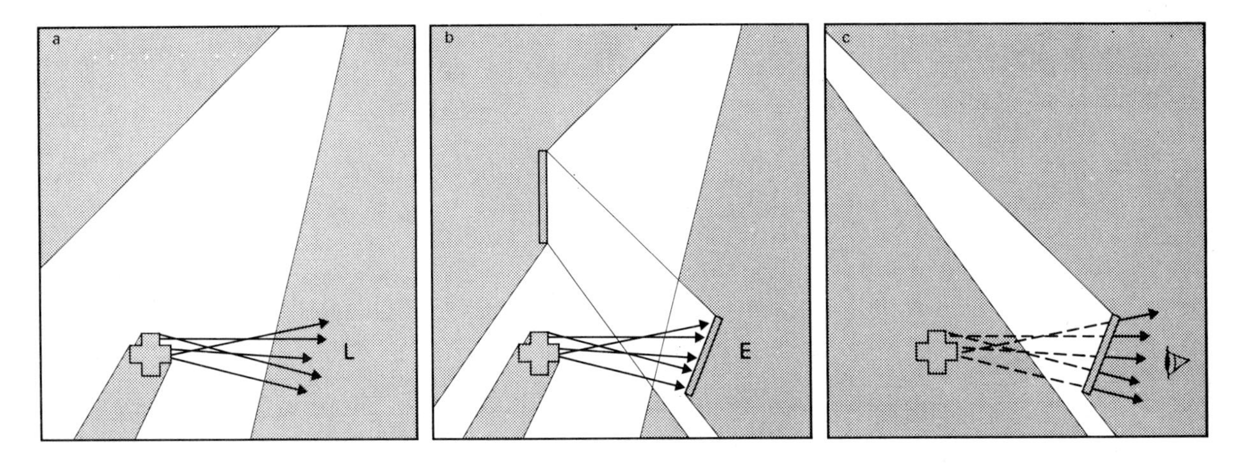

4.18 a–c *Principle of hologram*

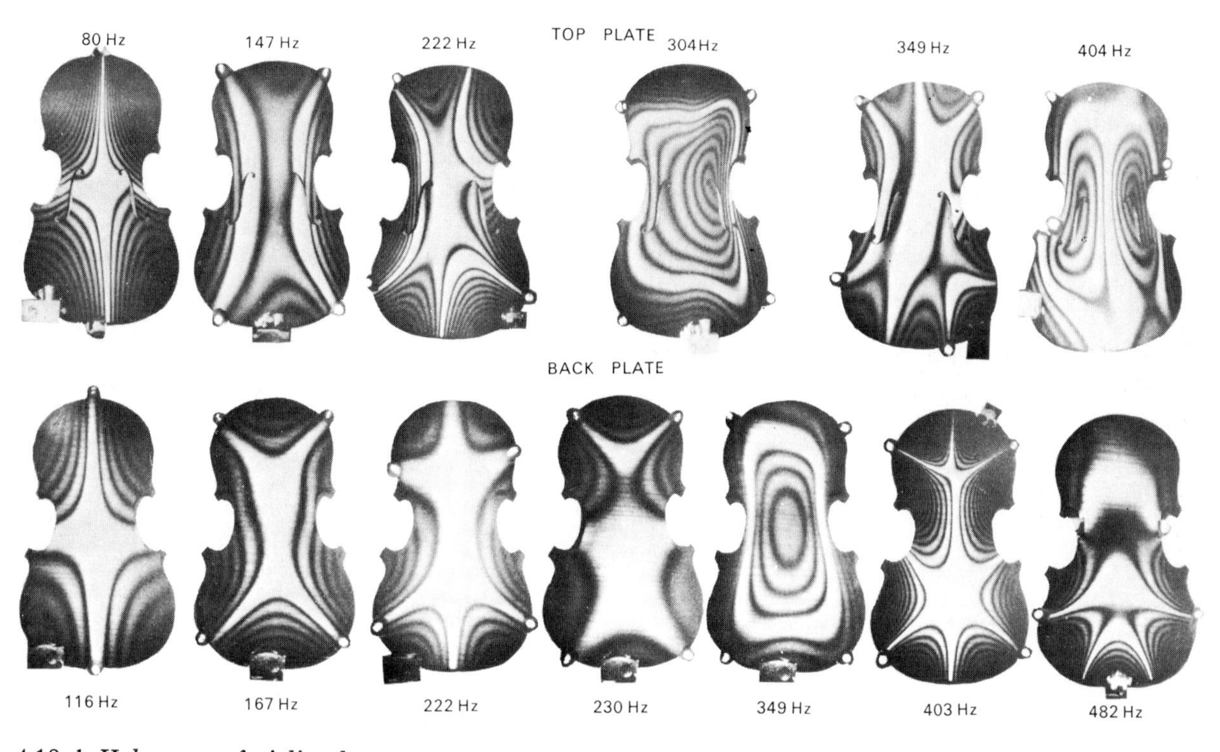

4.18 d *Holograms of violin plates*

to all points on the photographic plate, but each of the individual waves will have travelled a different distance. The result is that some waves will be in step with the beam reflected from the mirror and some will be out of step. Those in step will produce a bright patch of light and those out of step will produce no light, and hence a highly complicated and finely detailed pattern will result on the plate. Notice immediately that changes in the path length of half a wavelength of light (which might be about three-tenths of a millionth of a metre) could change the pattern from light to dark. Suppose now we take the developing plate and put it in a laser beam which exactly matches the one reflected from the mirror in the first stage (4.18 c).

The dark and light patches on the plate will turn this beam into a set of waves which are exactly like the waves that resulted from the interaction of the two sets of beams in the first stage. If we look into the developed plate, therefore, we receive in our eyes an identical set of waves to those we should have received if we had looked from position E in 4.18 b and hence we imagine that we can see a full three-dimensional image of the object behind the hologram (the photographic plate). This is a beautiful technique but how does it apply to violins? Suppose the object were a violin back or belly driven by a vibrator. Those parts which are not vibrating will form an image quite satisfactorily – but those parts which are vibrating even a very small amount (and remember we said path differences of less than half a wavelength matter) will record a different pattern on the hologram. If the original undisturbed back plate is placed in the position occupied by the image of the vibrating one the result is to produce outlines of the vibrating pattern – just like the sand on the Chladni plate – but with very small amounts of vibration. The so-called interfermetric hologram results from the wave pattern scattered by the still and moving plates differing enough to add or subtract and give light or dark bands. 4.18 d shows examples for a violin.

These and many other scientific approaches have led, not only to the production of instruments with predictable tone properties but also to the development by the Catgut Acoustical Society of a complete new range of instruments. The traditional violin, viola, 'cello and bass have quite different tone colours quite apart from their differences in pitch. The differences are used very skilfully by composers, especially in chamber music where small groups of four or five are involved. It would, however, be a fascinating experiment to produce a set of instruments with essentially the same tonal quality so that, for example, a particular melodic theme could be thrown from one to another as it goes out of range of any one instrument without an obvious change in tone quality. The goal aimed at by the Society was to produce a new octet of instruments, two of which are higher in pitch than the conventional violin and five are lower. The tone quality of all is modelled on that of the traditional violin which becomes the mezzo violin of the new family. The sequence is treble, soprano, mezzo, alto, tenor, baritone, bass and contrabass. The lowest note of each is either a fifth or a fourth (alternating) above that of the one below. 4.19 a shows their relative sizes, their pitches and their relationship to the size of tra-

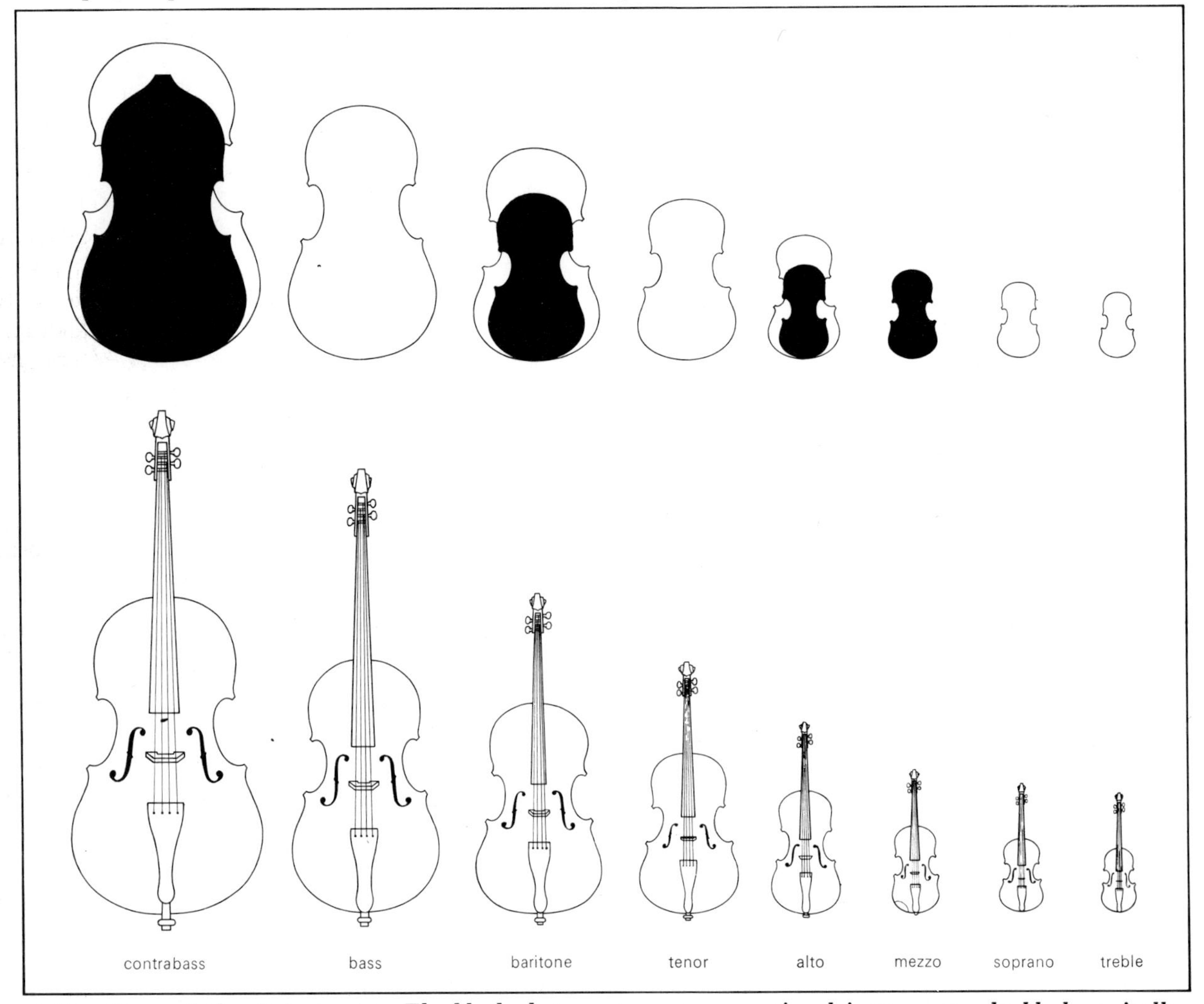

4.19 a *Sizes of the Catgut octet. The black shapes represent conventional instruments, double bass, 'cello, viola and violin*

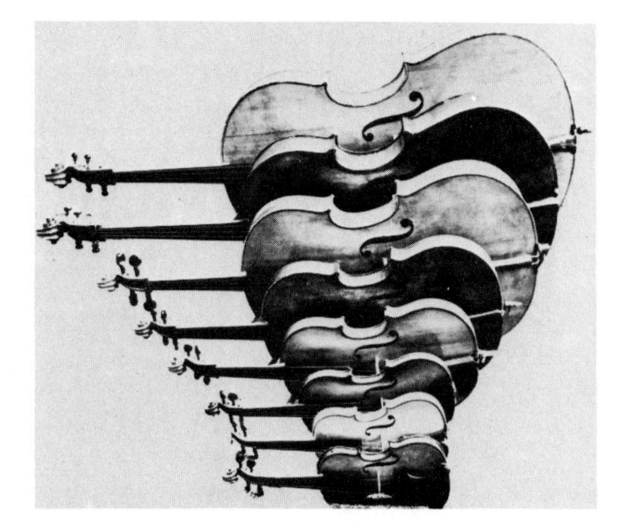

4.19 b *Photograph of the family*

ditional strings. 4.19 b shows the complete family.

The octet is now complete, music has been written for it and played, and undoubtedly it has unique and fascinating qualities which cannot help but have a lasting impact on the string family. 4.20 shows two members, the huge contrabass, and the alto, which is somewhat longer than a normal viola. It is normally played vertically, but here the performer demonstrates that with a good reach it can be fitted under the chin.

4.20 a *The alto*

4.20 b *The contrabass*

KEYBOARD INSTRUMENTS

Two families of keyboard instruments need concern us: the piano family– which started from attempts to mechanise the harp – and the organ family, which probably stems from attempts to permit one person to play a whole orchestra of instruments simultaneously.

The form of keyboard used, first in the virginal, clavichord and harpsichord and later in the pianoforte, has changed very little over the last 500 years, although the mechanism between the keys and the strings has gone through many evolutionary stages. Terminology and chronology of all the various early instruments in this family are both subjects of considerable argument and it would be inappropriate to discuss these matters here. We shall look briefly at three members of the family which have quite distinct mechanisms and tone qualities. The simplest is the clavichord. A set of parallel strings runs across in front of the player at right angles to

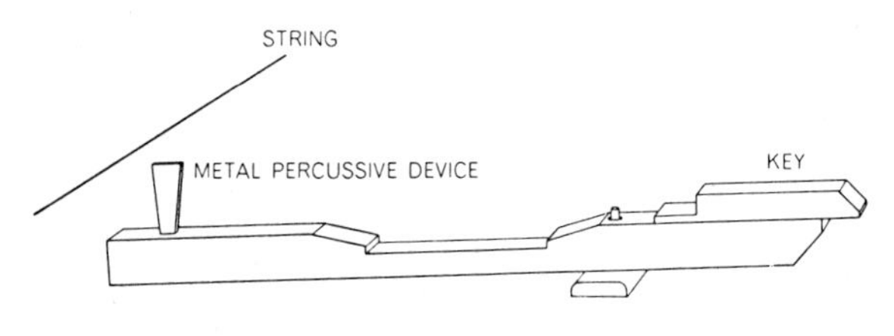

4.21 *Principle of the clavichord*

the keys. Each key is a simple lever which has a narrow metal 'hammer' – rather like a blunt chisel – fixed to its rear end so that when the front of the key is depressed the hammer rises, strikes a selected string and continues rising to form a bridge which divides the string into two parts (4.21). A strip of felt intertwined between the strings damps out the vibrations of one part and the other part goes on vibrating until the key is released. As soon as the hammer-bridge leaves the string the felt damps out the whole vibration. Several keys operate on each string, producing different notes according to the striking position. The instrument is compact, has the possibility of variations in loudness depending on the force with which the keys are struck, can produce vibrato if the key is 'wobbled' after the strike and while the hammer is acting as bridge, but suffers a great disadvantage, for most purposes, that the sound is very quiet indeed. The instrument in modern form has had a revival of popularity as a domestic instrument because it can be played in a modern flat without disturbing the neighbours! The principle can be demonstrated very easily with any stringed instrument, such as a guitar, violin, 'cello, etc., by striking one of the strings at various places with the back of a knife.

The virginal, spinet, harpsichord group differ mainly in size, shape and complexity, but all use the same principle of operation, which is plucking rather than striking. The key operates a 'jack' which lifts the string upwards on a piece of goose quill or similar material for a certain distance, after which the string is released and vibrates freely. When the key is released the mechanism allows the quill to pass the string without striking it and also lowers an individual damper on to the string to stop the vibration (4.22). The sound is louder than that of the clavichord, but the loudness cannot be varied since the string is drawn upwards the same distance before release whether the key is struck gently or violently and it is the *release* that sets the string vibrating. Many ingenious complications are known on large harpsichords such as multiple strings with stops so that the number of strings plucked can be changed to give a loudness control, multiple keyboards, 'lute' stops which bring in a separate set of jacks which pluck the strings at a different point and so produce a different quality of tone and so on. The pianoforte, however, must rank as one of the most brilliant pieces of mechanical craftsmanship. The 'action' of a modern grand piano is a masterpiece (4.23). Every key must 'feel' the same; every key must perform as nearly identically as possible; it must project the felt-covered hammer to strike the strings at a velocity which can be sensitively varied by changing the way in which the key is depressed; it must avoid the possibility of the hammer bouncing and hitting the string twice when the key is struck only once and yet must be sensitive enough to respond accurately and unfailingly when a key is struck many times in rapid succession as in a trill; it must provide damping action when the key is released. All this and much more is achieved by a collection of wooden rods and levers, metal pivots, felt pads

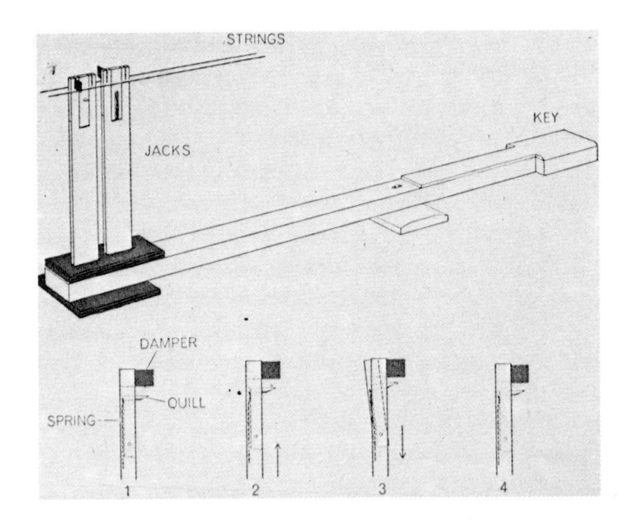

4.22 *Principle of the harpsichord*

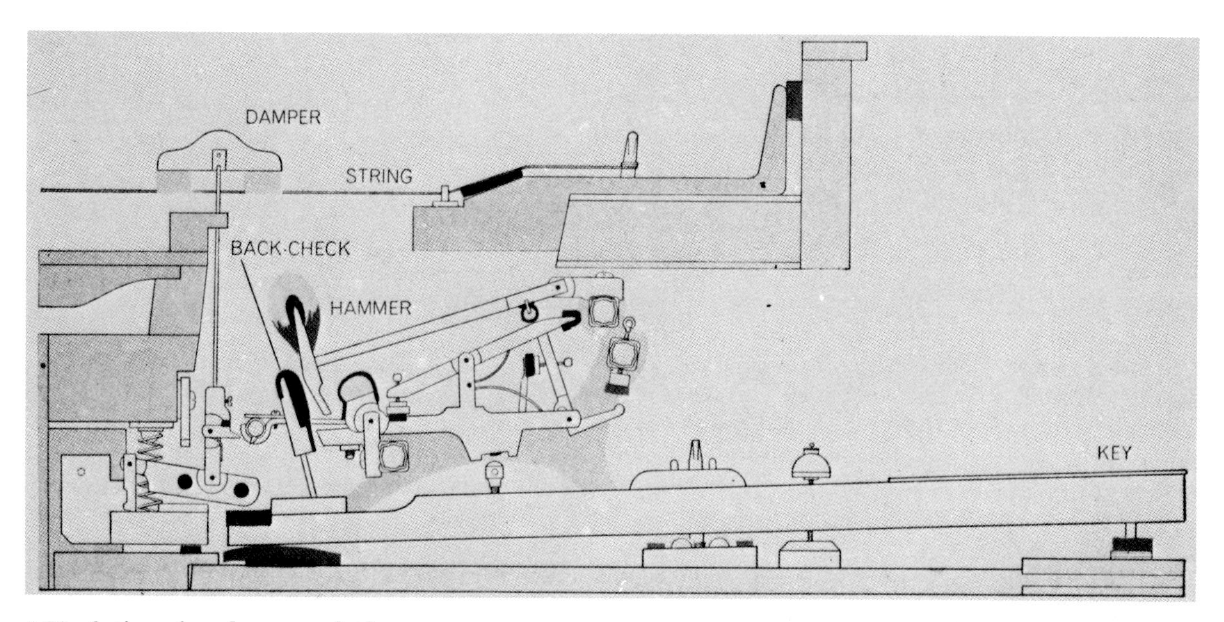

4.23 *Action of modern grand piano*

and leather strips which look like an inventor's nightmare and yet perform their required function beautifully.

As with so many topics, there is material for a whole book on the controversies that have raged round the secret of piano touch and tone. We will pick out just two points which arise naturally out of the discussions of earlier chapters.

Modern pianos have enormously strong, steel-framed sound-boards which amplify the vibrations of the strings, and impose formant characteristics in so doing. The great strength permits the use of quite stiff, thick steel strings stretched to very high tension, all of which leads to great increases in the volume of sound compared, for example, with the clavichord or harpsichord. These stiff strings, however, do not behave in quite the simple straightforward way that we might expect, and it turns out that the overtones are no longer exactly harmonic. The effect is small for the first few overtones, but by the time we reach the fifteenth or sixteenth it may be as much as a semitone sharp from that of a true sixteenth harmonic. The mistuning increases more rapidly with higher overtones and in some reported electronic analyses reaches nearly a fifth ($3\frac{1}{2}$ tones) sharp between the fortieth and fiftieth overtones. What effects arise from this? First, it has been shown by some workers that if we synthesise piano tone electronically and make the components strictly harmonic the tone sounds very different and – surprisingly enough – nowhere near as pleasant. Synthetic tone in which the inharmonicities of real piano strings are simulated immediately acquires the quality which musicians often describe as 'warmth'. We shall return to this point in the last chapter because it raises some fascinating questions about the way in which our brains work.

We have already mentioned in an earlier chapter (see, for example, 3.31) that upper partials may decay at rates quite different from those of the lower ones and these differential patterns of behaviour play a large part in enabling the player to have more control over the tone than would appear possible. This leads us to the second point of controversy. Between the two wars experiments were performed in which weights of different size were dropped on to piano keys and the resulting wave forms compared with that produced by virtuoso pianists; similar experiments were performed with the piano out of sight and listeners were asked to distinguish. But in most cases the experiments were confined to single notes taken out of context and it is not surprising that some scientists of the day proclaimed that the idea of variation of touch was a myth. In fact even with single notes some variations are possible for a skilled player other than merely altering the force with whichever key is depressed – for example, by decreasing the distance the hammer has to travel by partially depressing the key slowly and then accelerating rapidly. It is true, however, that the hammer is not in contact with the key when it strikes the string and therefore the only variable – however controlled – appears to be the velocity of the hammer just before it strikes the string. The clue appears, however, when one considers the playing of chords and rapid successions of notes in which all kinds of subtle variations in relative loudness of the harmonics and in the time of decay – controlled by both the individual damping on the keys and the sustaining pedals – can be achieved. The time of decay – because of the great difference in rates for different partials – has a considerable effect on quality. Undoubtedly this is only part of the story, but it will suffice to show once more how much the scientific and technological aspects of the problem become inextricably mingled with the artistic and aesthetic ones.

Now let us turn briefly to what has sometimes been described as the King of Instruments – the organ. The name covers everything from the simple short single-keyboard instrument – which is little more than a mouth-organ fitted with a bellows to supply the wind and keys to determine which metal reed vibrates, the reed organ or harmonium – to the great concert organs with four or five keyboards, pedals and a vast array of pipes producing as wide a variety of sound qualities as the instruments of a full orchestra. The basic principle of operation, however, is identical whatever the size. All the pipes, reeds or other sound-producing devices are grouped in three possible ways. First, all the pipes of whatever kind but of the same pitch are linked together and controlled by a given key; secondly all the pipes of the same quality of whatever pitch are linked together and controlled by a given stop,

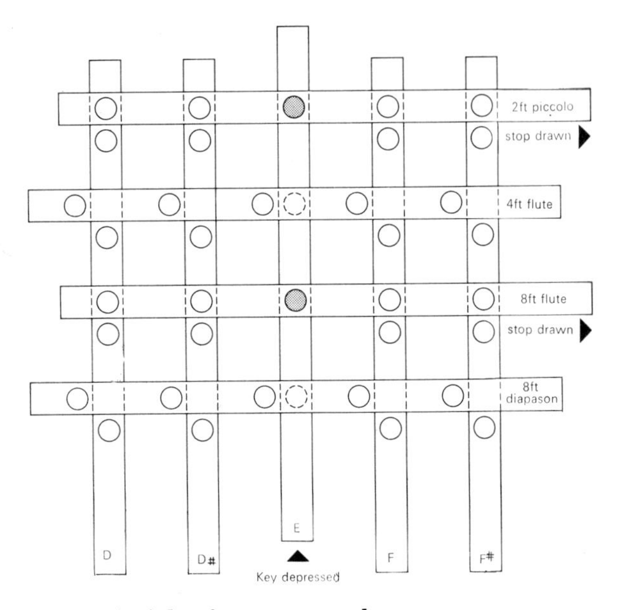

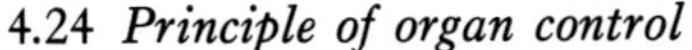

4.24 *Principle of organ control*

and finally groups of related keys (e.g. octaves on the same keyboard or notes of the same pitch on two or more keyboards) or pipes in given combinations are linked by couplers.

In order to play a single note on a single set or rank of pipes both the appropriate key must be pressed and the corresponding stop drawn (4.24). On a large organ the possible combinations are enormous. Each keyboard virtually controls a separate organ (except for the couplers which enable more than one organ to be played from the same keyboard). Even on quite small organs the pipes controlled by one keyboard are enclosed in a large box fitted with moving shutters like large venetian blinds to control the volume and to some extent the quality of sound. The 'swell' effect as the box is opened on a well-balanced set of flue and reed pipes is one of the glories of the large English organ; the box when closed tends to cut out more of the higher harmonics and hence as the sound increases in loudness there is also a characteristic increase in 'reediness'.

One of the interesting peculiarities of organs is the convention of describing the pitch of a rank of pipes in terms of the length of the longest pipe of a rank of simple open flue pipes of the same pitch. Thus a rank of pipes which sound at normal pitch, that is the same pitch as the corresponding notes on a piano, is described as '8-foot pitch'. A calculation on the assumption that the velocity of sound is about 1000 feet per second and that the pipe concerned is effectively open at both ends ('open diapason' – the basic tone of the organ) shows that the frequency is about 63 Hz, which is not far from that of the C two octaves below middle C which is usually the lowest note; 4-foot and 2-foot pitch naturally sound an octave and two octaves higher respectively and 16-foot and 32-foot pitch respectively an octave or two octaves lower. Various intermediate pitches also occur for use in producing tonal effects, e.g. 'twelfth' of $2\frac{2}{3}$-foot pitch, reinforcing the third harmonic, which occurs at the musical interval of a twelfth above 8-foot pitch.

It is important to remember that all keyboard instruments are really collections of instruments each of which is separately controlled – they belong to our family number one. Thus each pipe on the organ is only ever called on to produce one note, though it may do it alone or in concert. This means that the builder can design each pipe to give exactly the same quality of tone and it is unnecessary for the quality to change as one moves to upper and lower registers. On all other instruments, e.g. brass, woodwind, strings, compromises are necessary because the same vibrator amplifier and initiator are used for all notes. There is, however, an overall formant quality imposed by the surroundings of the organ – the swell box and the whole shape of the building. The organ should be voiced (i.e. the pipes finally adjusted to give the right quality of tone) when the organ is in position in the building in which it is to be used and the design of the organ and its surroundings are really inseparable. 4.25 a shows the organ in the hall of the Music Department at University College, Cardiff. The swell shutters can be seen in the centre. In 4.25 b can be seen the way in which the pipes fit special cavities in the wall; on the left are reed pipes and on the right wood and metal open diapasons.

4.25 a *The organ of the Music Department of University College, Cardiff*

4.25 b *The pipes fit special cavities in the wall*

Pipe organs are expensive to build and – because they contain a great deal of wood and leather – tend to be sensitive to changes in moisture and temperature and so can be expensive to maintain. Nowadays increasing use is being made of electronic organs and they undoubtedly can be made to give an acceptable sound at lower initial and maintenance costs. When first introduced they were not at all comparable in tone with a pipe organ. In the very early models the builders concentrated on producing a steady wave form which matched that produced by a pipe organ. This is not difficult and they succeeded by blending pure tones produced by oscillators or rotating generators. Unfortunately, although they could produce steady wave forms identical with the steady wave form of, say, an open diapason, a flute or a clarinet stop – when actually played all three sounded 'electronic'. The problem was that in aiming for the steady-rate wave form the result was far too steady; they had ignored the transients, the difference in rate of build-up of the various partials, the 'breathiness' at the beginning of each note – all the elements which in some terms might be thought of as 'imperfections' which in fact turn the clinically pure wave form of a simple tone generator into the warmth and reality of a 'real' instrument. Modern electronic organs reproduce all these transients and other variations tolerably well and there are models available which, when properly matched to the building in which they are housed and with multiple loudspeakers carefully sited to give a realistic sound distribution, are very difficult to distinguish from pipe organs except by real experts.

There are many different systems of tone generation – again a subject for a separate volume – but we will give a glimpse of four just to indicate the possibilities. The simplest – though not necessarily the cheapest by a long way – is to provide a large number of oscillators, each producing pure tones of a given pitch, and then by a complex switching system to arrange that when a given key is depressed and a given stop is drawn the right tones are selected, mixed and fed to a main amplifier and loudspeaker to produce the desired pitch and quality.

The second popular system is to use basic tone generators that are oscillators but which produce a wave form itself very rich indeed in harmonics. The resultant sound is harsh and reedy, but the desired changes in quality corresponding to the various stops can be produced by filtering the basic wave form – i.e. using a simple spectrum-shaping circuit which changes the harmonic mixtures.

The other two methods to be described have mechanical systems as the source of the basic tones and this gives them, among other features, the advantage that they cannot go out of tune and indeed the overall pitch can be raised or lowered if necessary to match some other instrument.

The Compton electrone system uses rotary tone generators which produce tiny changes in electrical capacity that can be converted into a changing voltage which is then amplified for use. In the simplest such organ there are twelve identical tone generators, one for each note of the equal-tempered scale C C♯ D D♯ . . . etc. Each tone generator produces seven notes an octave apart by having seven 'tracks' of varying shape as the stator and seven rings of 'teeth' on the rotor which move over the metal tracks and produce variations in capacity. 4.26 shows a typical tone generator. This one generator produces seven Cs an octave apart. Each generator is driven by means of a pulley and belt from

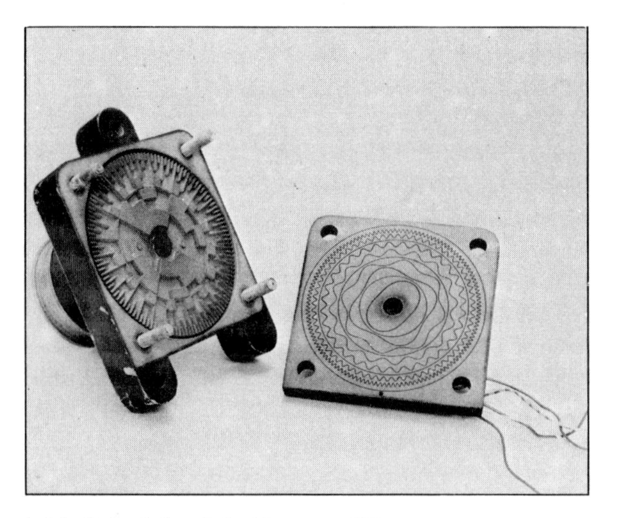

4.26 *Principle of the Compton Electrone tone generator*

a common motor shaft, but the diameters of the driving pulleys are different for each generator so that the frequencies have the right ratios to give the required succession of notes. The second generator thus produces seven C♯s, the next seven Ds and so on. Twelve generators thus produce eighty-four basic notes which can then be selected and mixed, have transients superimposed on them and so on. More complex generators themselves produce complex wave forms and, as can easily be seen, the possible combinations of this scheme are very large.

The Hammond organ also uses rotating generators, but this time they are electromagnetic. Rotating metal discs with shaped edges generate varying currents in coils wrapped round magnets fixed near to the edge of the discs. Again by a combination of variations in the speed of rotation from group to group and by varying the shape of the disc both in terms of number of high points or teeth and in their shape, basic sets of related frequencies can be produced from which pitch and quality selections can be made (see 4.27).

Great ingenuity in circuit design and in other ways has brought electronic organs to an established place as serious musical instruments and no doubt the future will bring more developments as electronic technology progresses. An interesting philosophical problem soon arises, however, and that is whether one should regard electronic instruments as attempts to copy existing tonal qualities – which is clearly what they attempted to do at first – or whether one should regard them as totally new instruments. I think that electronic organs in particular suffered in the early days because many musicians insisted on regarding them as inadequate substitutes for 'real' organs. Nowadays the flood of electronic instruments is such that musicians are beginning to explore them as new instruments in their own right.

NEW MECHANICAL INSTRUMENTS

Not many entirely new musical instruments based on purely mechanical principles have appeared in the last fifty years or so. Very recently a toy came onto the market which consists of about a metre or so of corrugated plastic pipe open at both ends which, when whirled round like a sling, emits ghostly notes whose pitch can be varied over four or five harmonics by varying the speed of rotation. The source of the excitation is the air passing through the tube as a result of centrifugal action interacting with the corrugations. But this is hardly a serious instrument. The 'steel-band' developed in the West Indies from home-made originals created from old oil drums has become a serious instrument with a remarkable range of tonal qualities, especially when played in relatively large groups. The sound

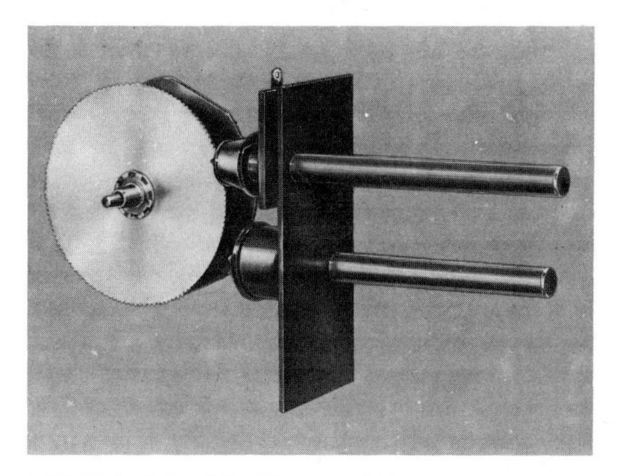

4.27 *Principle of the Hammond Organ tone generator*

is produced by creating separate vibrating sections on a plate, each of which is tuned by a combination of size, shape and thickness to a different note. Creation of this instrument – whether from old oil drums or from specially produced materials – is a highly skilled empirical art. The principle is not completely unrelated to that of the musical saw, in which you will remember that the metal must be bent into an 'S' curve and it is effectively the centre section of the 'S' supported freely by the springiness of the two end curves that gives the extremely resonant and loud tone. The end of the steel drum is first beaten inwards to be concave and then the individual regions are beaten upwards to become convex. Each vibrating region is thus itself convex but is surrounded by a 'moat' which acts like the curved ends of the saw.

By far the most interesting new development from a purely scientific point of view, however, is the creation by François and Bernard Baschet and Jacques Lasry of a new family of instruments, 'Les Structures Sonores'. Their instruments have all the characteristics that were identified in the first three chapters and one can see quite clearly the part that is played by each element. The primary vibrator (see 4.28) which produces the basic sound is a length of rod – in the earlier versions merely 6 mm screwed brass

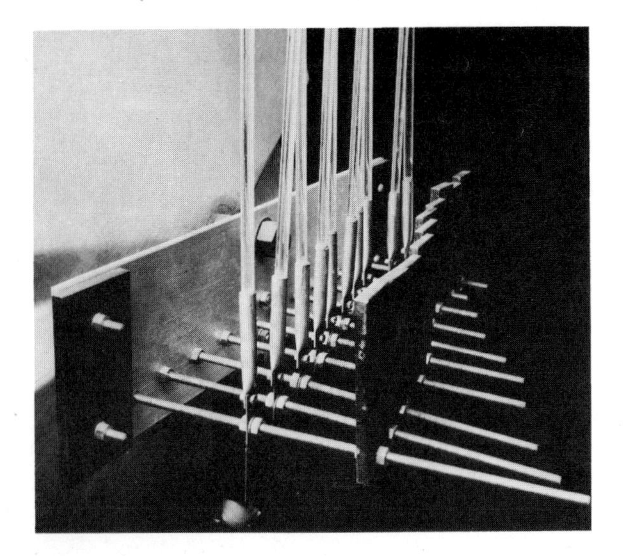

4.28 *Vibrator unit of Structure Sonore*

4.29 *Bladder and string in an engraving by Hogarth*

rod. One end is screwed firmly into a thick strip of metal and a weight at the other end provides additional inertia to make the sound persist longer and also provides a means of tuning by adjusting its position on the rod. Just as with a tiny fork, the sounds produced are not very loud and a mechanical amplifier is needed. The vibrations are transmitted by the metal strip either to a sheet-metal horn via wires or strings, to large cylinders which give resonance, or in some examples to inflated plastic air cushions. Inflated cushions or balloons make very good amplifiers though they impose very curious formants. The idea is really quite an old one and in one of Hogarth's drawings (4.29) can be seen a 'bladder and string' which is virtually a one-string fiddle using a pig's bladder as the amplifier.

In the French instruments additional formant qualities are sometimes added by collections of steel wires attached at one end to the main steel bar and hanging freely at the other. The instruments have been designed with a remarkable eye for visual appeal as well as musical quality. The remaining factor is the method of starting up the vibration. In some the bars are merely struck – but in others glass rods some half metre or so in length are fitted at right angles to the vibrators at an adjustable position, and *longitudinal* vibrations of these glass rods induced by stroking with wet fingers (as in the musical glasses) become transverse vibrations of the brass rods. (In some versions the brass rods are vertical and the glass rods horizontal for ease of playing.) The resultant tone is quite unique. The transients are fairly slow and organ-like, but the steady-state quality varies from string-like to brass-like. Perhaps one of its most characteristic features is that the quality varies very much with amplitude and the same instrument can purr a quiet melody or rasp out trombone-like snarls.

COMPUTERS AND SYNTHESISERS

Really high-speed, large-capacity digital computers have only been available for about ten years and yet already we regard them as indispensable elements in scientific programmes. The ever-increasing speed of progress in technology makes it extremely difficult to keep pace with all the applications that appear and the whole pattern of work may change in a matter of a year or two. Nowhere has this been more true than in the applications of computer techniques to the production of new sounds.

My first – somewhat crude – experience of computer music dates back perhaps fifteen years or so when a visit of a party of laymen to a computer laboratory always included the computer playing 'God save the Queen' not always precisely in tune and sometimes with rather peculiar phrasing! This was really in the category of a trick and all that was being done was to write a programme in such a way that pulses on the way to a store would follow each other at audio frequencies and, if fed to a loudspeaker, would produce harsh but nevertheless identifiable musical tones. The lack of precise tuning might occur because the rates at which the pulses could be produced were not infinitely variable and so some of the frequencies were 'rounded off' to the nearest practicable number. Although very crude, such tricks form the basis of various systems which have been developed for producing sounds of high quality and of infinitely varied tone colour, and about seven years ago we were already listening to an IBM computer singing 'Daisy, Daisy' (with an American accent).

The principle of using a digital computer to produce musical sounds is very simple. The computer itself is neither more nor less than a device which can store a great many digits in each of a large number of locations and with tremendous rapidity can transfer digits from one place to another according to the instructions, or programme, fed to it. All the complex mathematical operations it performs are really made up of simple transfers; the secret of the computer's power lies in the unbelievable speed at which it works and the vast number of digits that it can keep 'in its head' at one time. The essential step in producing computer music is a device known as a digital-to-analogue converter, or DAC for short. The DAC takes a set of numbers which

4.30 *Les Structures Sonores*

pour out of the computer and turns them into a varying electric current which, in turn, can be fed to a loudspeaker directly to be converted into sound waves or, more usually, recorded on magnetic tape as an intermediate step.

The DAC first produces a sequence of identical pulses at very short time intervals; in a typical example there might be 40,000 pulses every second. If this succession of exactly equal pulses were fed to a loudspeaker it is doubtful whether the diaphragm of even the best available would be able to move back and forth at anything like the required speed, and even if it did, the frequency of the waves produced, which would have a fundamental at 40,000 Hz, is above the limit of audibility for human ears. The DAC, however, does not keep the pulses identical in height. It reads the output of the computer and makes the height of each successive pulse correspond to each successive number emerging. 4.31 a shows the sequence of pulses that would emerge without computer control. Suppose now that the sequence of numbers emerging from the computer and controlling the height of successive pulses is 20, 19, 18, 17, 16, 15, 14, 13, 12, 11, 10, 9, 8, 7, 6, 5, 4, 3, 2, 1, 0, 1, 2, 3 . . . 19, 20, 19, 18 . . . 1, 0, 1, 2, . . . etc. The result would like 4.31 b. A quick sum shows that the pulses rise to their maximum height of 20 units 1000 times every second. Now imagine this fed to a loudspeaker. The successively decreasing and increasing pulses would gradually move the diaphragm and it would follow, not the individual pulses, but the overall *envelope* of the pulses which in this case is the shape shown in 4.31 c. We have synthesised a triangular wave form with a fundamental frequency of 1000 Hz which is right in the middle of the audio range. Clearly by persuading the computer to feed out highly complicated sequences of numbers, highly complicated wave forms can be produced and, in principle at least, any sound can be synthesised. All the features we have studied can be built in; transients, vibrato, rise and decay, formants, all can be included. Indeed, again in principle, manipulation of the wave form is extremely easy. Suppose, for example, that we wish to produce a tone containing eight harmonics in given amplitude ratios with a rise from zero to full amplitude in one-tenth of a second and immediate decay to zero over a period of one second, together with a light vibrato at a rate of six 'wobbles' per second – what do we have to do? The computer could *separately* generate the sequence of numbers giving the wave form of the fundamental and store it; then it could generate the sequence of numbers for the second harmonic and store it; and so on until all eight are ready and of the right relative amplitude. Then the set of numbers corresponding to the vibrato amplitude could be generated and stored, and finally the set of numbers giving the envelope amplitude at each 1/40,000 second. Now the computer could add the first number in each of the stores corresponding to the harmonics together, multiply it by the number of the vibrato store and by the number in the envelope store, and feed the result to the DAC and so on. This is, of course, a gross over-simplification and in practice many thousands of such operations are done simultaneously and not in sequence, but it should illustrate the principle.

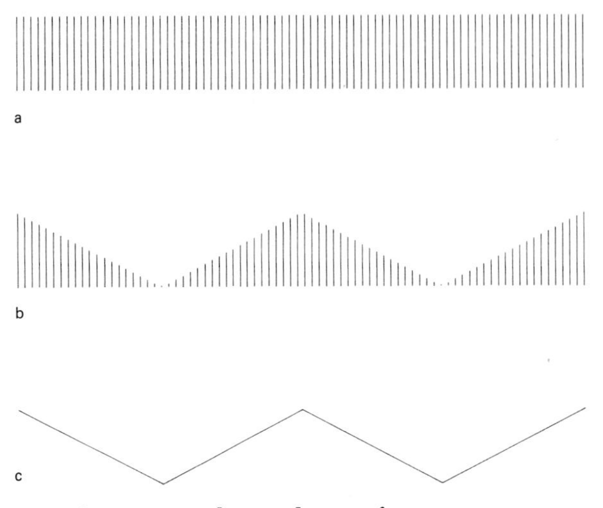

4.31 *Computer pulses and wave forms*

What are the practical problems? Perhaps it will not be surprising to hear that, in order to approach the kind of subtle variations and complications of tone that occur in real combinations of instruments the number of digits that has to be programmed and generated every 1/40,000 second is just too great even for our ultra-high-speed computers. We cannot therefore create music in 'real time', but it may take ten, a hundred or more seconds to perform the necessary computations to produce the final computer wave form for one second of music and so instead of listening as one goes along the output must be accumulated slowly on tape and then, when enough has been recorded, the tape can be played at normal speed so that one can hear the result.

Many ingenious ideas have been thrown up for use in this field. A whole 'orchestra' of ready-made sounds of many different pitches and qualities can be pre-calculated and stored so that the final creation of the music merely involves calling forth the appropriate numbers, adding them together and feeding the DAC. Such a procedure is much faster but is in turn limited by the demands on storage space in the computer for the 'orchestra'.

This kind of consideration, among others, has led to a move away from the somewhat cumbersome procedure of literally creating the waves at 1/40,000 second intervals towards the use of real wave forms that can be created, added, multiplied and generally combined to order in the device that has come to be known as a 'synthesiser'. In computer terms, this is merely a move from digital to analogue computing. The difference

4.32 *A clock as an analogue device: the angles of 30° and* $91\frac{1}{4}°$ *are instantly recognised and interpreted as 3-05*

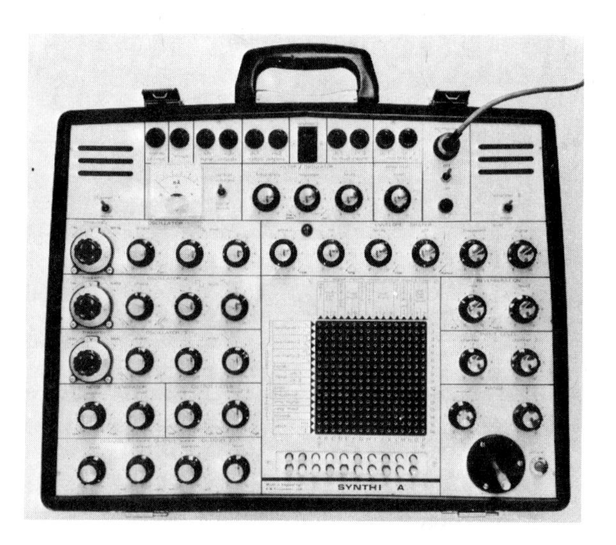

4.33 *A small synthesiser*

can be demonstrated most easily by comparing the type of clock that is often seen nowadays in stations and elsewhere – the digital clock – with the old-fashioned kind. In the new type the actual number is displayed and the digit representing the minutes flips over once every minute. In the conventional clock, however, we use as an analogue of the numbers representing the time the *angle* at which two arms are pointing. Indeed on the watch I am wearing at the moment there are no figures at all – merely twelve dots at 0°, 30°, 60°, 90°, 120°, 150°, 180°, 210°, 240°, 270°, 300° and 330°. I know what time it is by converting the angular position of the two hands with respect to the dots into numbers (4.32). I am doing the reverse of the DAC; when I tell the time I am performing analogue-to-digital conversion!

Many other analogues can be used, and in the synthesiser we use the electrical *voltage* to represent all the digits associated with a wave; in adding, subtracting, multiplying voltages the end result is the same as adding, subtracting and multiplying numbers, but is far quicker and does not involve such enormous storage problems.

In order to illustrate how a synthesiser works I shall use one of the smaller types now available as an example. The 'suitcase' version is shown in 4.33. There are four kinds of basic unit and they can be interconnected in a wide variety of

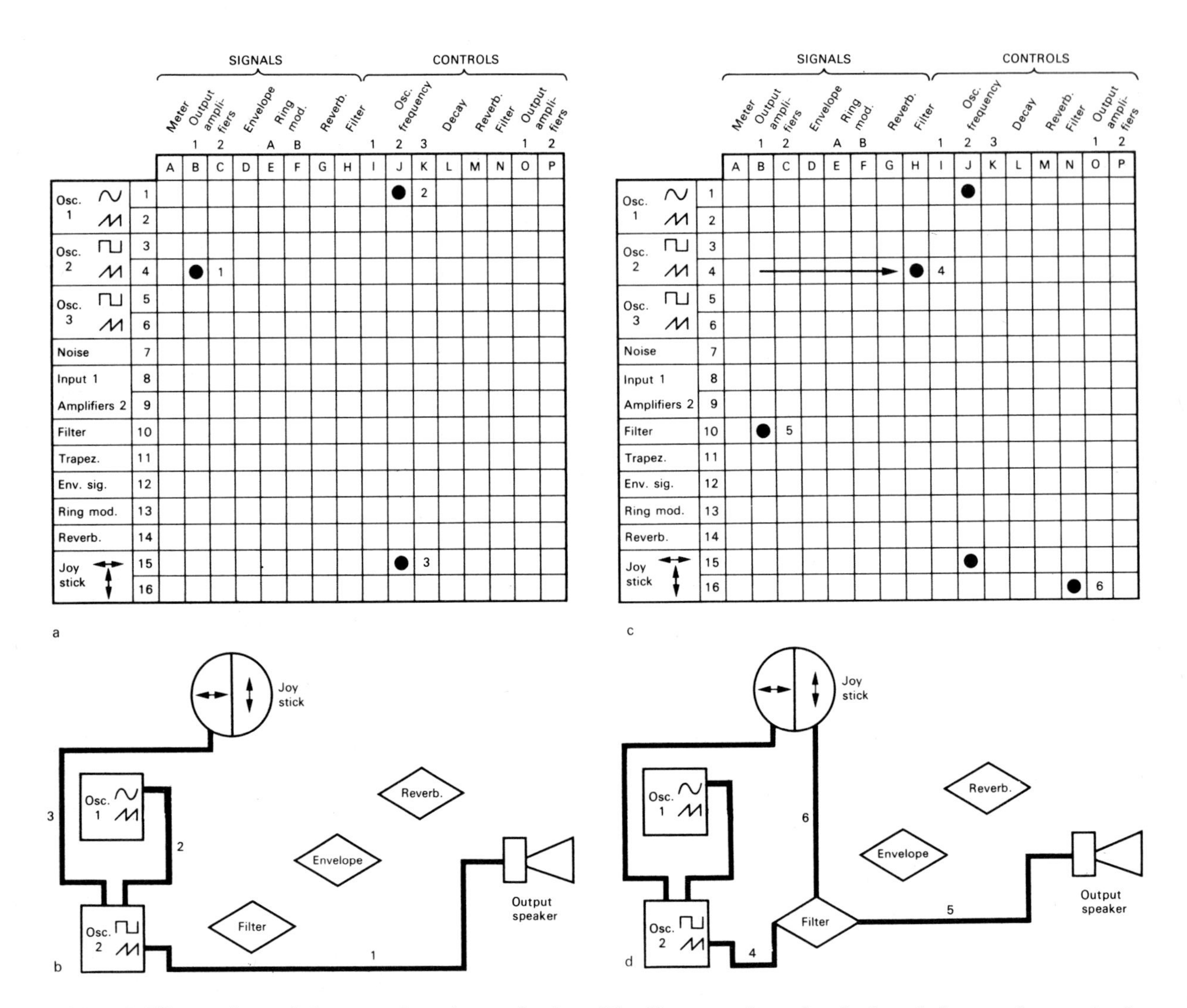

4.34 a–d *Illustrations of the operation of a synthesiser. The diagrams show the pin-board that can be seen in the illustration of the instrument on page 117*

ways by means of the matrix- or pin-board which is really the nerve-centre of the whole device. The basic units are source units or generators, such as the three oscillators and the noise generator; treatment units, such as the filter which modifies the spectrum, the envelope shaper which varies the amplitude with time; controllers, such as the joy-stick or keyboard which can be used to control various aspects of the final wave form as will be seen later; and interconnecting devices, such as the ring modulators and the pin-board itself. The real secret of the flexibility of the synthesisers, of which this is a typical example, is that each unit can be controlled by a variable voltage and each unit itself produces a varying voltage and so all kinds of interconnections are possible. This last point may be a bit obscure unless I immediately give an illustration. Oscillator number one can be set to produce a pure tone and the left-hand knob controls the frequency between 1 Hz and 10,000 Hz. This knob can be manually set. Equally well, however, the frequency can be controlled by applying a voltage to oscillator number 1 in such a way that when the voltage is low the pitch is low and when high the pitch is high. Oscillator number 2 works the same way – so if we set oscillator number 2 at say 7 Hz and use its output to control the frequency of oscillator number 1 which is set manually at 440 Hz the result is 440 Hz with a 7-Hz 'wobble' or vibrato! Similarly if the voltage control of number 1 is now hitched to the keyboard – which merely produces a successively higher voltage as one presses keys going up the scale – the pitch of 440 Hz can be changed to something else and, by adjusting the levels of the controls carefully, the keyboard can be used to make oscillator number 1 give notes which correspond exactly to the normal scale on a piano; an interesting point is that it is equally possible to make the notes correspond to any other scale by suitable adjustments.

However, we are jumping ahead too quickly and, for those who are interested in understanding the flexibility of the system and how it is used, the next few paragraphs give a fairly detailed description; those who are not so closely interested could skip on to page 120, last paragraph.

Opposite above are diagrammatic views of the pin-board; we shall go through a series of settings and the black dots on these diagrams indicate the insertion of a pin and the numbers alongside indicate the order of the operations. Below are block diagrams of some of the units and give the linkages corresponding to the pin settings on the diagram. The numbers alongside each link identify the order in which the links are established and correspond to the pin numbers above.

The conventions in the block diagram are that sources of signals are square, treatment units are diamond-shaped and control units are circular. Signals (i.e. varying voltages that could be used either as sounds or to control something else) leave units from the right and, if intended to become sounds, enter at the left; control voltages enter at the top and leave at the bottom.

The sequence of operations will now be described in words and it is hoped that this, together with the diagrams, will make the operation clear.

We begin with diagrams 4.34 a and b.

Operation 1. Insert a pin at 4B. The signal from the saw-tooth output of oscillator number 2 is connected to the output loudspeaker and we hear a steady rather reedy note whose frequency can be controlled by tuning the frequency dial on oscillator number 2.

Operation 2. Insert a pin at 1J. The signal from the sine-wave output of oscillator number 1 now controls the frequency of oscillator number 2 and if the amplitude of 1 is set to be rather small and its frequency to say 6–7 Hz a convincing vibrato is imposed on the signal from oscillator 2.

Operation 3. Insert pin 15J. Now left–right movement of the joy-stick will also control the frequency of oscillator 2 and so we can adjust its mean frequency easily to anything we like while maintaining the vibrato as well.

Now turn to 4.34 c and d.

Operation 4. Remove the pin from 4B and place it at 4H. The output is now going to the filter instead of to the loudspeaker.

Operation 5. Insert pin at 10B. The output from the filter now feeds the loudspeaker and we again have a sound, but it has been modified in quality by the formant characteristic of the filter.

Operation 6. Insert pin at 16N. The formant characteristic of the filter can now be changed by up-and-down movement of the joystick.

At this stage we have a continuous reedy note with vibrato whose mean frequency can be moved up and down by left–right joystick movements and whose quality can be made more or less reedy by up-down movements of the joystick.

Now turn to 4.35 a and b overleaf.

Operation 7. Remove pin from 10B and insert at 10D. The output of the filter now goes to the envelope shaper instead of the loudspeaker.

Operation 8. Add pin at 12B. The output from the shaper now goes to the loudspeaker, but no sound emerges unless the 'trigger' button is pressed and then we obtain a single note whose envelope characteristic – i.e. rise, fall, duration, etc. – can be set by adjusting the control knobs on the envelope shaper.

Now turn to 4.35 c and d.

Operation 9. Remove pin from 12B and insert at 12G. The output from the shaper now goes to the reverberation unit instead of the loudspeaker.

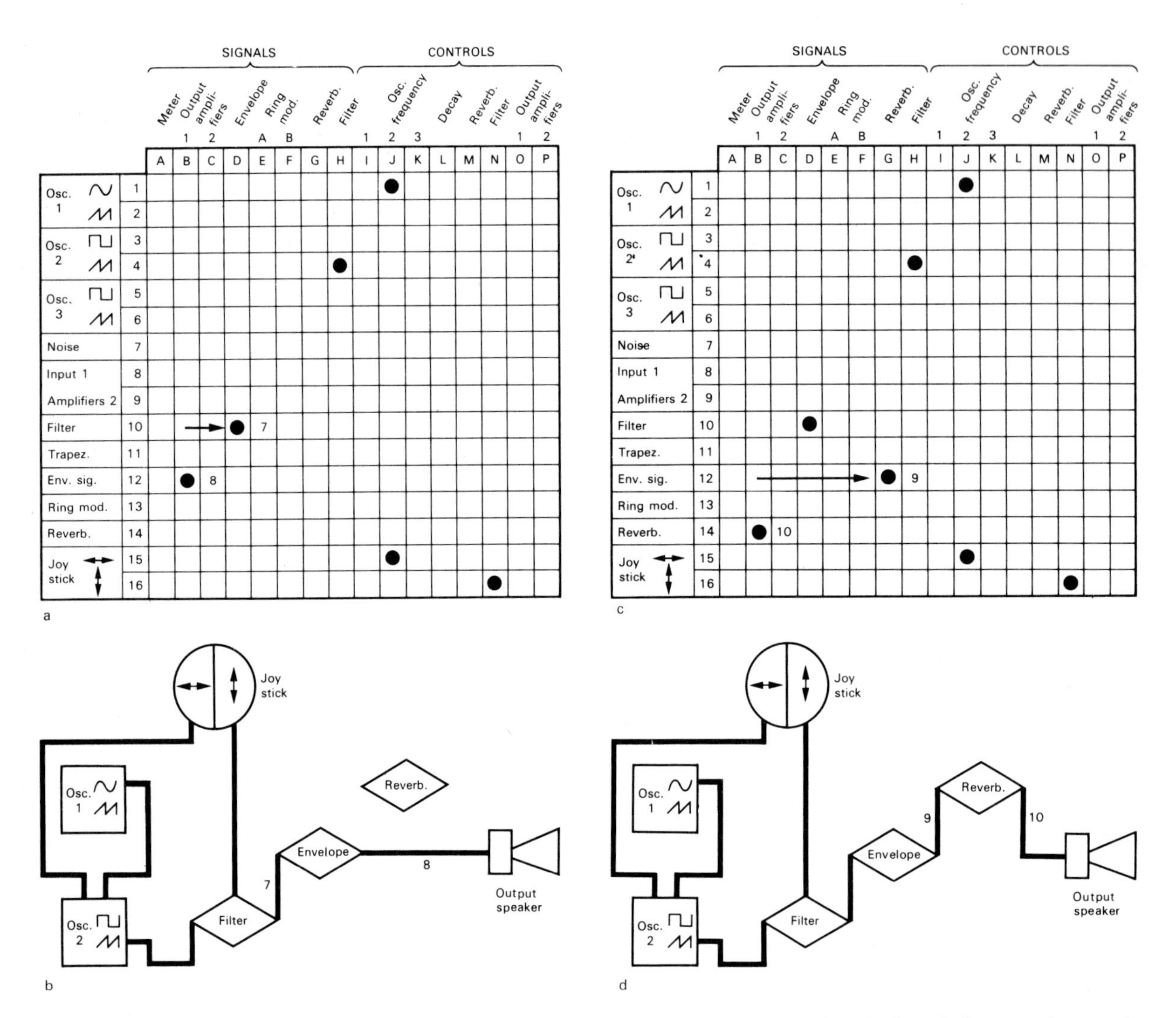

4.35 a–d *Illustrations of the operation of a synthesiser. The diagrams show the pin-board that can be seen in the illustration of the instrument on page 117*

Operation 10. Insert pin at 14B. The output of the reverberation unit now goes to the loudspeaker and we can control the amount of reverberation added to each note by the controls on the reverberation unit.

We have now arrived at a fairly simple 'patch pattern' of seven pins which enables us to produce single notes of variable pitch, frequency, quality and envelope shape with vibrato and reverberation. We could use this to play a melody or to record a single part of a complicated composition. Far more complex interconnections are used in practice, but this sequence should indicate the principle underlying the system.

4.36 shows a very much larger version – the Synthi 100 – as installed in the Physics Department at University College, Cardiff. The principle of operation is exactly the same; two large matrix boards are provided, one of which links signals, and the other treatments, and there are many more individual units. The section at the right-hand side is an extremely useful addition that does not occur on the small synthesiser; it is a memory unit.

When a portion of a melody or a single part

of a composition has been prepared it can be played and remembered by this unit. At any time then a touch on the 'start' button plays the melody back. What is stored, however, is not the actual wave form, but the control voltages involved. This is where the great economy arises relative to digital computers. For example, the information needed in the synthesiser to produce a continuous tone of a certain frequency is merely one set of numbers representing that frequency and a second set indicating the required loudness level. The computer, however, needs a new set of numbers indicating the pressure in the wave 40,000 times per second!

4.36 *Large synthesiser in the Electronic Music Studio, University College, Cardiff*

The synthesiser memory and sequencer unit can store up to three different melodies and will play them back together when required so that sound textures can be built up and the relative timings are easy to achieve. Wrong notes can easily be corrected and the various parameters of notes can be modified during play-back because, remember, only the control signals are stored, not the actual wave forms. Short sections of compositions are prepared in this way and then transferred to tape to build up the finished work. The economy is obvious and the logical conclusion is that a combination of a synthesiser such as this with a quite small computer, which only needs to calculate sets of numbers corresponding to control voltages, is far more powerful for musical composition than an enormous but entirely digital computer. The same system may be used for producing all kinds of controlled wave forms for psycho-acoustic experiments of the type to be described in chapter six and so once more there is a fruitful link between research in physics and creative music.

I have described this synthesiser because it is the one I have used and which we have in our laboratories. There are many others – some commercially available and some unique designs built for particular uses and to do a particular job. There are many variations in the electronic techniques used in the units and in the switching devices used to control them, but the underlying philosophy is much the same in all of them and it would be out of place here to pursue the details.

What of the future of electronic music? In the past there has been a tendency to use electronic sounds as background music for space films or science fiction plays and this can have unfortunate consequences – rather like those facing an actor who becomes 'typecast' in a particular role and finds that his audiences always think of his original character even when he is playing an entirely different part. The phase is already beginning to pass, however, and synthesisers are becoming sufficiently common to attract the attention of many serious composers. My personal view is that the synthesiser should be welcomed as a totally new instrument which increases immeasurably the range and control that the composer has of tone qualities, but that it is complementary to conventional instruments rather than competitive, in the same way that the voices and the orchestra are complementary in opera. Indeed some of the most exciting compositions to date are those in which electronic music is introduced as a third component or dimension together with voices and instruments.

A FINAL WORD

Modern technology is also playing an important part in the study of the human voice – which is, I suppose, itself a magnificent example of economical and efficient musical instrument design. We mentioned at the end of chapter three the use of X-ray techniques for studying the shapes of the voice-producing cavities and of course many examples of techniques of electronic analysis could be given. Professor Schouten and his colleagues at the Institute of Perception in Eindhoven have produced some beautiful demonstrations using 'gating' circuits which allow only a specifically chosen fraction of a word to pass through to an analysing device and the aural effects are sometimes quite startling. For example, a word like 'Music' might be passed through an electronic gate which opens and closes just long enough to allow a recognisable bit of the word to be heard; then the moment at which the gate opens is delayed a fraction of a second at a time. At first one hears 'Mmm', then 'Mee', then 'eeoo' then 'ooze' and so on. Analysis in this way draws attention to the important factors and features in speech.

We have no room for a long discussion of the various ways of synthesising speech which are important both in drawing attention again to the really vital parts on which recognition depends and also as possible ways of communicating with computers in the future.

Just as with electronic music there seem to be two principal avenues of approach. One is to synthesise the speech wave forms from first principles, using the computer technique mentioned in the last section. The other is more like the synthesiser technique in which voices are built up by electronic analogues of the various parts of the human vocal system – devices for producing basic tones corresponding to those produced by the vocal chords which rise and fall to give the intonation, formant shapers which impose vowel sounds, noise generators that can supply the 'f's', 's's', etc., and envelope circuits which can produce the 'p's', 'b's', etc. Exciting results are already being achieved and the future I am sure holds some fascinating possibilities in this particular branch of technology.

SIGNOR GRUNTINELLI.
PLAYING ON A NEW INSTRUMENT (CALL'D A SWINETTA)
Pub. Accord. to Act, Jan. 1. 1774 by M. Darly 39 Strand

The concert, French tapestry, about 1500

ON THE WAY TO THE EAR

EVERYBODY MUST BE SOMEWHERE

In the first three chapters we considered various more or less scientific aspects of the problem of creating sound waves in the air and in the last chapter we paused to think about the consequences for instrument makers and musicians. Now we must get back to our central theme and see what happens as the waves created by instruments or voices start out on their journey to the ear.

First let us think about a purely imaginary experiment. Two little men (5.1 a) are suspended, perhaps by will-power, in the atmosphere but miles away from any other solid objects. If they tried to hold a normal conversation they would soon discover that it would be very difficult if they were more than about 10 metres apart. Why is this? Look at 5.1 b in which we have sketched in the way in which the pressure waves created by each man spread out – and remember that this ought to be a three-dimensional picture since the 'wave surfaces' (that is, the surfaces indicating where a particular pressure change has got to at a particular time) are spherical. Now think back to chapter one for a moment and you may recall that we said that waves are a means of transferring energy from one point to another. How effectively is man A transferring energy to man B? Suppose man A expends a certain amount of energy in setting up a wave – he might shout or clap his hands. The wave set up in the air spreads out more or less spherically and the total energy is spread out over the ever-increasing surface of the sphere. Suppose the second man B is 10 metres away – the sphere will have a radius of 10 metres by the time it reaches him and the area of its surface will be 1257 square metres! (The area of the surface of a sphere is $4\pi r^2$.) Suppose that man B has enormous ears and is able to funnel down to his ear drums a 25-square-centimetre sample of the wave – and this is certainly far more than the average. This represents a fraction 25/12,570,000 (since there are 10,000 square centimetres in a square metre) of the surface of the sphere. Just about one part in half a million of the surface is used by man B, and since the energy was spread uniformly this means that only one part in half a million of the energy given out by man A will be received by man B. What an inefficient way of communicating! No wonder they have difficulty in holding a conversation. One further point: the man B does not change the size of his ears and so, if he moves further away, the total energy is spread even more thinly over the growing surface of the sphere and he picks up an even smaller fraction. In fact if he goes twice as far away he will pick up a quarter of the energy, so the fall-off is quite rapid.

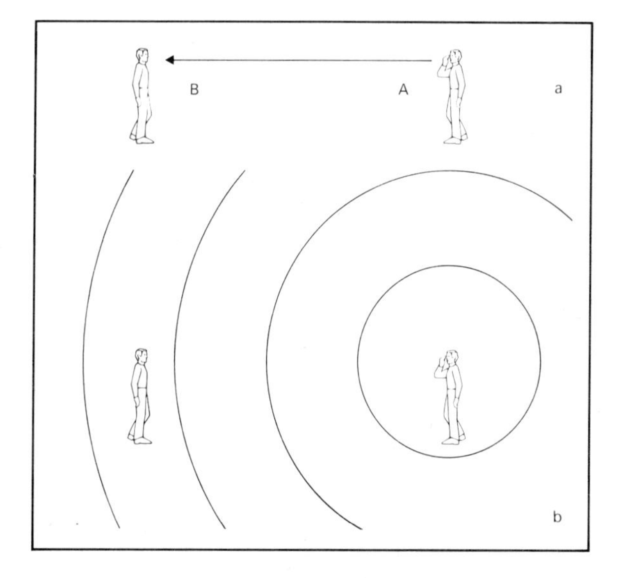

5.1 *(a) Conversation in space. (b) The waves involved*

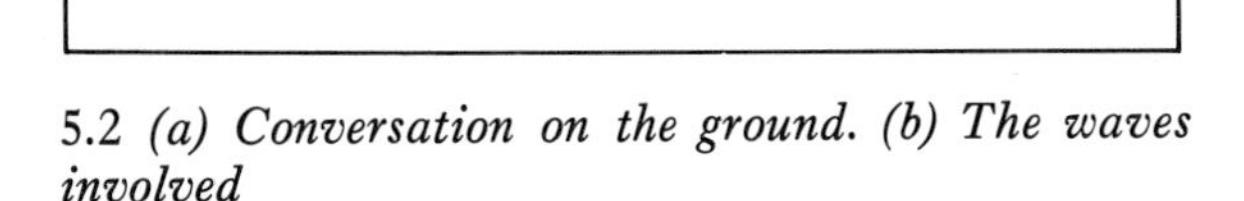

5.2 *(a) Conversation on the ground. (b) The waves involved*

Fortunately this situation could hardly arise. As the title of this section says, 'everybody must be somewhere', and, at the very least, we usually have our feet on the ground. 5.2 a shows the effect of the ground, assuming for the moment that the waves bounce off the ground as though reflected by a mirror. Man B now receives two lots of waves and the effect is as though he is receiving sound direct from man A and also from an upside-down reflection of him in the ground surface (5.2 b). In practice the reflection of the waves – even if the surface of the ground is smooth and hard like a concrete playground – is not as perfect as the reflection of light from a mirror and it becomes quite fuzzy if the ground is very rough, as for example a ploughed field. Nevertheless this comparison with reflection of light gives us a useful picture to start us off on the right lines as long as we do not push the parallel too far. Suppose now we give man A a wall. 5.3 a shows that there are now three paths by which the waves can travel from A to B in addition to the direct one and 5.3 b illustrates how they correspond to three extra sources of sound. There may well be other reasons for a street orator having his back to a wall, but it makes good acoustic sense if he wants to be heard by his audience!

As soon as we add another flat surface parallel to one of the first ones – a ceiling or a rear wall – the waves can bounce back and forth for a long time and each rebound corresponds to another reflected man. In fact we end up with (theoretically) an infinite series of reflections and so the level of sound heard goes up enormously. 5.4 a and b illustrate this. The effect is just like that produced when an object is placed between two mirrors which are parallel to each other. 5.5 shows a photograph of a doll placed between two parallel mirrors and photographed round the edge of one of them.

Now we can finish off the story by completely closing in both men with parallel walls and ceiling – they are now in fact in a room and we shall have three sets of infinite numbers of reflections corresponding to the three pairs of sur-

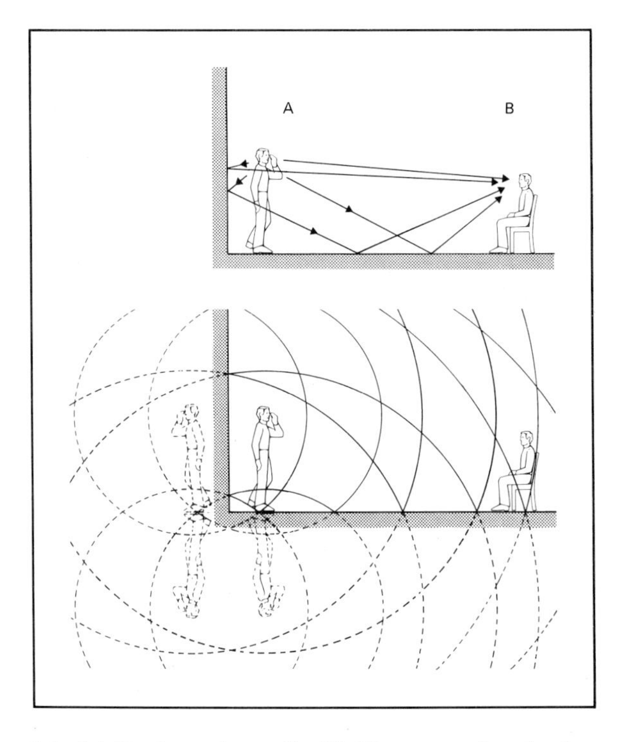

5.3 *(a) Back to the wall. (b) The waves involved*

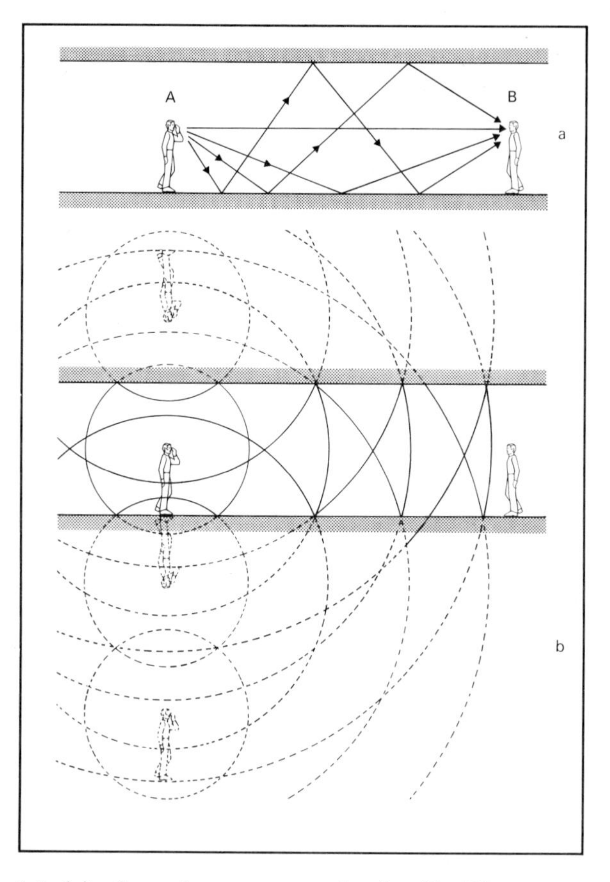

5.4 *(a) A roof over your head. (b) The waves involved*

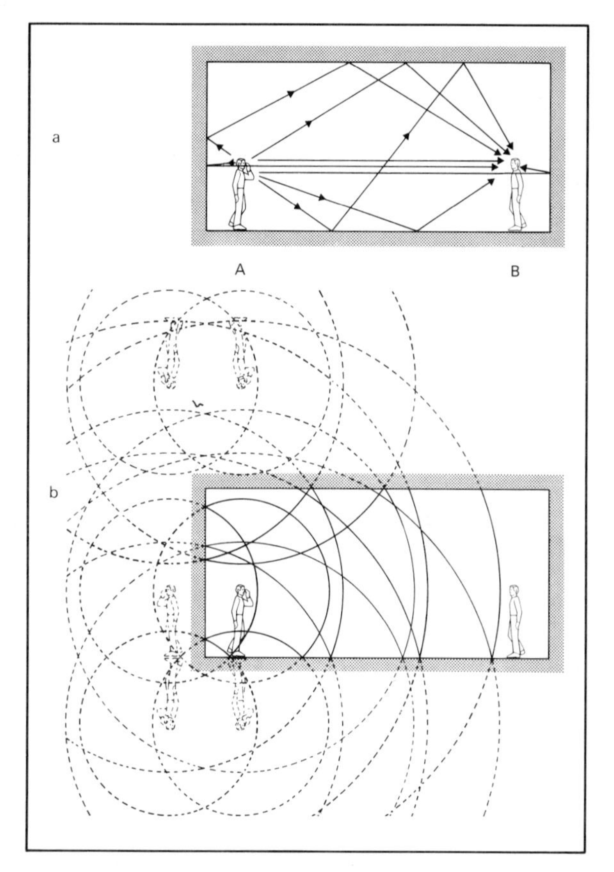

5.6 *(a) Conversation in a room. (b) The waves involved*

5.5 *The 'infinite repetition' effect with parallel mirrors*

faces making up the room (5.6). Does this help A and B to hold their conversation? If the surfaces are all very hard and smooth then it will *hinder* rather than help, because the waves will go on bouncing back and forth inside the room. Now, if you will again cast your mind to chapter one, you will remember that we emphasised the importance of the time taken for waves to travel from one place to another. In this example it will take time for the waves to bounce back and forth, and the bigger the room the longer will the waves take on each trip. A sharp sound like a single click is thus drawn out in time and could last indefinitely if the walls were perfect reflectors. Thus each syllable pronounced by A will be drawn out in time and will overlap all the successive ones; in such a room B would find the conversation quite unintelligible. The

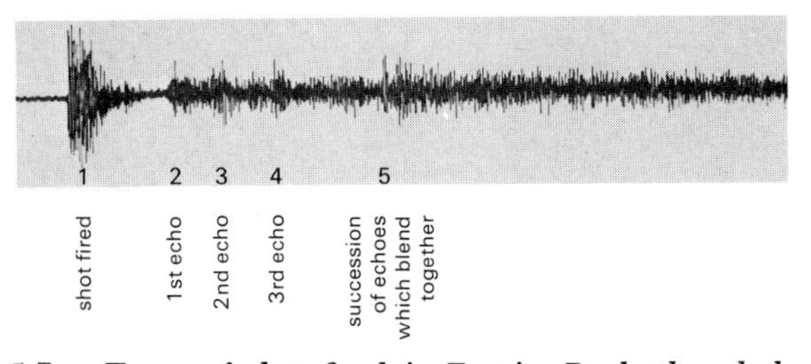

5.7 a *Trace of shot fired in Empire Pool: the whole trace as shown lasts about half a second. The first echo corresponds roughly to a reflection from a surface 35–40 feet away and is probably the roof; the second and third are about 45 and 60 feet and represent reflections from side or near end walls. The continuous succession starts with one from a surface about 100 feet away and after that echoes follow rapidly and blend into each other*

closest approach to this in everyday life is in an indoor swimming pool where the water surface acts as a hard, smooth reflector and all the other surfaces are hard, smooth glass or tiles. 5.7 a is the oscillograph trace of a recording of a single pistol shot made in the Empire Pool, Cardiff (5.7 b) – which incidentally has surprisingly good acoustic properties and is much less noisy than some others – but nevertheless you can see clearly the first reflections arriving at the microphone; after that the many different reflections all tend to become blurred into a general sound which gradually dies away. If the walls were *perfect* reflectors it would never die away and each tiny sound would go on for ever and after a very short time the noise level built up would become quite unbearable. Fortunately there are always, in any real room, surfaces which do not reflect perfectly; they absorb varying amounts of the sound energy and prevent this intolerable situation.

SINGING IN THE BATH

You can probably already see that we are in something of a quandary. If we have no walls at all it is very difficult to hold a conversation, and yet if we surround our two little men with perfectly reflecting walls an impossible situation develops in which the slightest sound goes on for ever and conversation becomes equally impossible. By careful choice of the placing of the walls and the nature of their surfaces, however, it is possible to arrive at a suitable compromise whereby we can increase the sound level enough for normal conversation without making it so high that everything becomes blurred. It should be clear, however, that the compromise chosen will almost certainly depend on the exact use that is to be made of a room. A luxuriously furnished room with thick pile carpet, rich velvet curtains, well-padded easy chairs tends to be a quiet place and in such a room one can almost feel the silence descend as one enters. The slightest noise is quickly absorbed and the room is highly suitable for an intimate chat at close range. The bathroom in most houses, however, has hard tiled surfaces and little in the way of soft materials, and hence any sound introduced tends to persist. Singing in the bath is therefore a very satisfactory experience (for the singer!) because the many reflections permit quite a high volume of sound to be reached relatively easily and the lengthening of each sound by the multiple reflections tends to blur out wobbles and slight catches in the breath and makes it relatively easy to create enjoyable sounds. In fact we are introducing yet another formant characteristic – the formant of the room. The room has a particular shape and size and there will be some frequencies that are amplified more than others – especially if the room is small and of simple shape. In general, larger rooms or rooms of very complex shape will have such complex resonant patterns that the amplification tends to be much more uniform. It would be useful if we could express all these ideas in a slightly more scientific way, and to see how this can be done we must look back at some delightfully simple experiments done by a Harvard physicist, W. C. Sabine, about 50 years ago. Sabine was concerned about the properties of a particular lecture room and began to do some simple experiments. He quickly realised that one of the first things to be done was to achieve the right balance between too much reflection – which would give overlapping and consequent blurring of syllables – and too little, which would not help the lecturer to be heard at the back of the

5.7 b *The shot being fired*

theatre. He proposed a quantity which could easily be measured called the 'time of reverberation'. Roughly this is the time taken for a loud sound to die away to nothing. A crude way to measure it is to fire a pistol shot and start a stop-watch simultaneously – then stop the watch when the sound can no longer be heard. In a highly reverberant room – a swimming pool for example – this time might be as much as 6 or 7 seconds (in the Empire pool mentioned above it is 3–4 seconds); in a small, softly furnished lounge it might be as little as half a second. (Strictly speaking – in order to have a precise definition – the time of reverberation is specified as the time taken for a sound to decay to the minimum audible level from a level a million times higher – but in practice the rough definition given above will give results that do not differ by more than a fifth of a second or so from those of the more precise one.)

Sabine's experiments were to measure the time of reverberation when the room was empty and all the seat cushions were taken out; he then brought back various lengths of seat cushion and found, as he expected, that the time became shorter as more sound-absorbent material was brought in. Seat cushions vary from theatre to theatre and one could hardly use them as standards, so Sabine then had the ingenious thought that the most perfect absorber possible is an open window – the sound goes out and none is reflected at all. He therefore did a second series of experiments in which the time of reverberation was measured in terms of the number of square feet of open window at any time. In this way he could arrive at a relationship between the absorption of the cushions and the absorption of the open window, and from then on all his results could be expressed in terms of 'o.w.u.', or *open window units*. For example, 100 square feet of wooden panelling might turn out to have an absorption of 25 o.w.u. – that is, they produce the same lowering of the reverberation time as 25 square feet of open window. In this case each square foot of panel has 25 per cent of the effectiveness of open window and we might say that its absorption coefficient is 25 per cent, or 0.25.

It is probably clear, since time is involved, that a big room is likely to have a longer time of reverberation – even with the same amount of absorption present – because it will take longer for the sound waves to travel from wall to wall. In fact Sabine devised a very simple formula which is still used today. It simply says that if the volume of a room in cubic feet is V and the total absorption present in equivalent square feet of open window is A then the time of reverberation is

$$T = \frac{V}{20A} \text{ approximately.}$$

We do not intend to delve into mathematical arguments, but perhaps two simple sums might help to show how this works. Suppose we have a swimming pool which is 100 feet long, 40 feet wide and 25 feet high. Its volume is exactly 100,000 cubic feet. The total surface area is 15,000 square feet and all the surfaces, including the water, will have very low absorption coefficients – perhaps 5 per cent, in which case

$$T = \frac{100{,}000}{20 \times \frac{5 \times 15{,}000}{100}} = 6{\cdot}6 \text{ seconds.}$$

Alternatively consider a lounge which is 20 feet by 15 feet by 10 feet high. Its volume is thus 3000 cubic feet and the total surface area will be 1300 square feet. Suppose that the average absorption coefficient of all the surfaces is 25 per cent, then the time of reverberation will be

$$T = \frac{3000}{20 \times \frac{25 \times 1300}{100}} = 0.46 \text{ second approx.}$$

This ties in reasonably well with the times given earlier in the section.

The absorption coefficients of most common materials are now known; we also know that an adult has an equivalent absorption of just under 5 o.w.u. and so it is possible to make a fairly good estimate in advance of building or furnishing a hall or theatre what its reverberation will be with and without an audience.

YOU CAN'T PLEASE EVERYBODY

This is all very well, but what do we do then? What *should* the time be for 'good acoustics'? This is the point at which we have to leave behind simple scientific experiment and enter the realms of judgement. If we know exactly what the room is to be used for and it is only to be used in one way, then it is not too difficult to decide what T should be – although even then tastes vary. Suppose, for example, the room is to be used for chamber-music concerts: one might think that there would be one specific time that would be right. It turns out, however, that it depends on who is asked! The performers will have one view – and different players may well have different views as some prefer the room to be 'lively' and others prefer a 'drier' acoustic. In other words, some like a longer time of reverberation and others shorter. The audiences may well have different views, and almost certainly a BBC recording engineer who is trying to site his microphones in such a way as to convey an accurate impression of the sound to the listeners will have different ideas again. Add to this the fact that very few rooms are used for one purpose and a room which is fine for a small committee meeting, for example, may be quite impossible for a choir rehearsal, and one begins to understand the problems of the architectural-acoustic consultant. The science is fairly well established, but the human factor is all-important. 5.8 gives a graph of the range of values of T which are reasonable for different sizes of rooms, but can only act as a very rough guide.

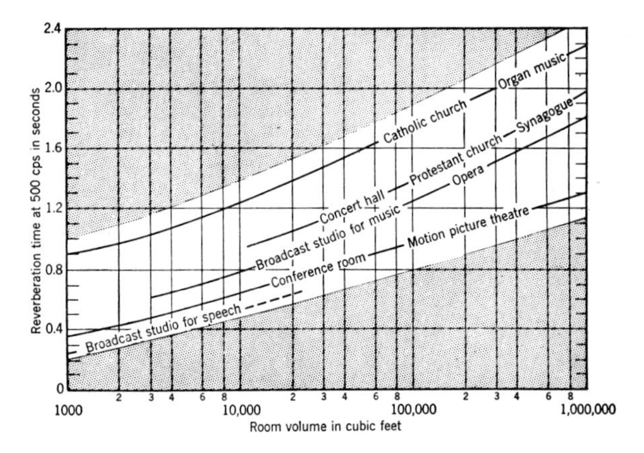

5.8 *The approximate relationship between volume and the acceptable reverberation time for normal use*

The mention of multiple usage of rooms raises a further question. Is the time of reverberation the only acoustic factor that matters? Before answering the question I want to ask you to do a mental experiment which may help you to answer the question for yourself. Imagine a room which is fairly long and narrow. Exactly half the room – one end wall and the half of the walls, floor and ceiling next to it – is covered with heavily absorbent material and the other half is covered with hard glazed tiles. Now let us bring back our little men A and B. Suppose A sings near the absorbent end and B listens from the tiled end, and then they change places and A sings again. Calculation of T by the formula we used above gives one value and it will be the same throughout the experiment. Do you think that A and B will notice a difference when they change over?

I hope your answer is 'yes', because they certainly will. When A is at the absorbent end the waves which reach the room surfaces first will be heavily absorbed and only a fraction of the waves he produces will get down to the tiled end (5.9 a). The images of A in the floor, ceiling and side walls will fade out very rapidly and he will feel as though he is singing in a dry room. B listening, on the other hand, will feel that he is in a lively room and any slight movement he makes – a cough or a shuffling of his chair – will reverberate quite well. He will be rather puzzled at the quiet sound from A. Now when they change places (5.9 b), waves can bounce back and forth at the 'lively' end without much loss and A will feel that the time of reverberation is very much higher. B, now listening at the dead end, will feel that he is in an absorbent room and will hear A's voice with added resonance carry over quite well. In the early days of broadcasting when microphones were scarce and were not at all directional this latter configuration was often used with the microphone in the place of the listener B. This of course is an extreme case, but I hope it demonstrates to you that the distribution of the absorbent and reflecting surfaces in relation to the sources of sound and the listeners is quite important.

How could these principles be applied in practice? Suppose we are designing a hall which is to be the size of the swimming pool quoted earlier, but clearly we do not want it to behave acoustically in the same way. Suppose also that for most of the time the sounds will be made on a stage at one end and listened to by an audience in the middle and at the far end (5.10 a).

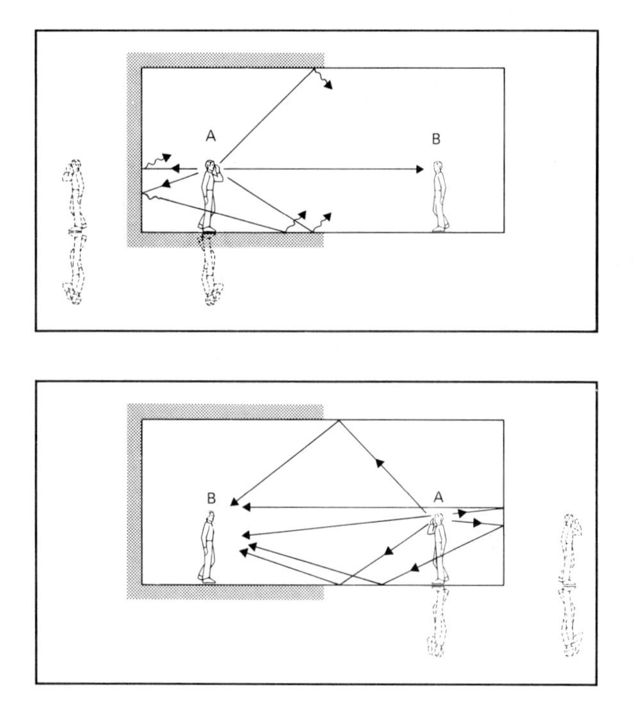

5.9 a and b *A room which is half reverberant and half dead*

The first problem that springs to mind is that the back wall is about 75 feet from the first row of the audience. A sound made on the stage will pass over the listeners on the first row, travel 75 feet to the back again to reach the first row for a second time; 150 feet takes about 0.15 second for sound to traverse and so the first row hears each sound at least twice with about 1/6 second interval – quite apart from all the other reflections. This time interval is too long for a direct echo and could become very irritating. Clearly then we should put absorbent material on the rear wall and possibly towards the rear of the side walls and ceiling. The walls near the stage need to be hard to help build up

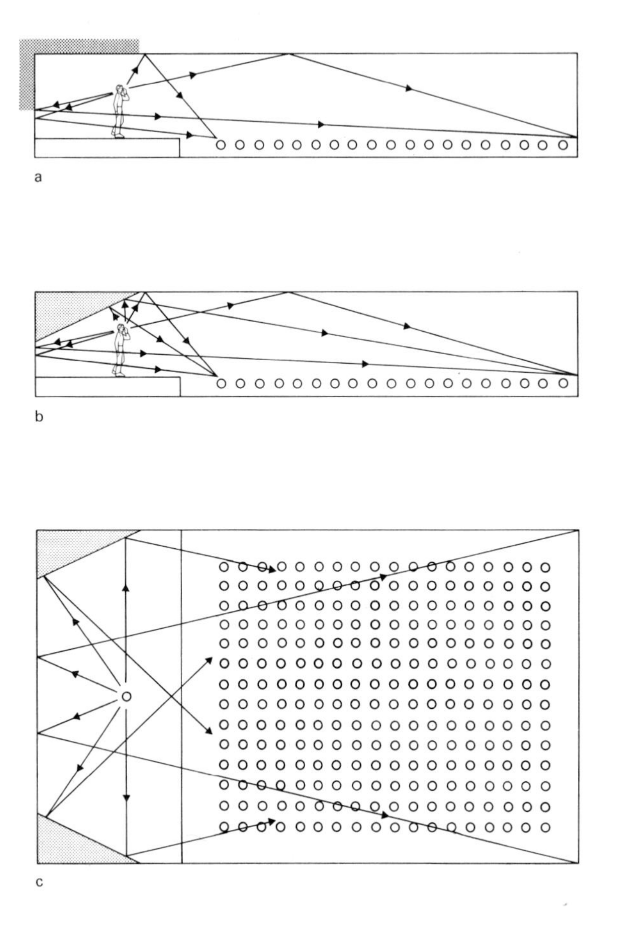

5.10 *(a) The shaded position is not helpful to the speaker. (b) The flare reinforces the speaker. (c) Plan view showing how side flares also help*

a loud sound which can be heard all over the hall. But look again at 5.10 a. The shaded parts of the stage wall and ceiling are not being used. Suppose, however, we introduced a sloping section of flare (5.10 b). Instead of wasting these parts they are now contributing very usefully to the reinforcement of the sound in the room. The same argument applies to the side walls shown in 5.10 c.

It looks from this sort of argument as though the ideal arrangement might be to surround the performer with a parabolic reflector – like that of a searchlight – and so use every bit of the sound waves he is producing to flood the audience and then have absorbent at the back to prevent echo. Once again, however, we find the need for compromise. Reflectors work both ways and every tiny sound made by the audience – coughs, sniffs, rustles, etc. – would be focused down on to the poor performer and he would experience an impossible noise level for good performance. It would resemble having one's ear to a giant shell and the 'roar of the sea' experienced would be overwhelming.

There are many more points to be taken into account, but this should suffice to show that the largest problem is in determining what the various users really want rather than how a customer, who knows exactly how he wants to use the hall and what acoustic properties he needs, can make sure before the hall is built that he gets what he wants.

SCIENCE AND ACOUSTIC DESIGN

In the first place it is quite easy to do calculations of the volume of the proposed room and of its total absorption in terms of areas and of the natures of the various surfaces. Sabine's formula then gives a prediction of the reverberation time and this can then be adjusted to suit the desired performance. Special acoustic-absorbent material such as spray-on fibrous finishes, acoustic tiles ($1\text{–}1\frac{1}{2}$ inches thick, rather like Weetabix or Shredded Wheat in appearance inside and usually with holes or cracks in the surface) or slatted wood panels with fibre-glass behind, can be used to vary the time. One complication, however, is that the absorption properties of all materials differ at different frequencies and it is all too easy to find that a hall has good properties for high-pitched sounds but has a long reverberation time for the bass note. Such a hall is sometimes described as having a 'boom'. It is important, therefore, to choose material which gives the right balance over the whole important range of frequencies.

The next problem of course is to decide on the distribution of the various materials and to test the proposed shape to ensure that the desired sounds are uniformly distributed to all parts of the audience.

Many devices have been used from time to

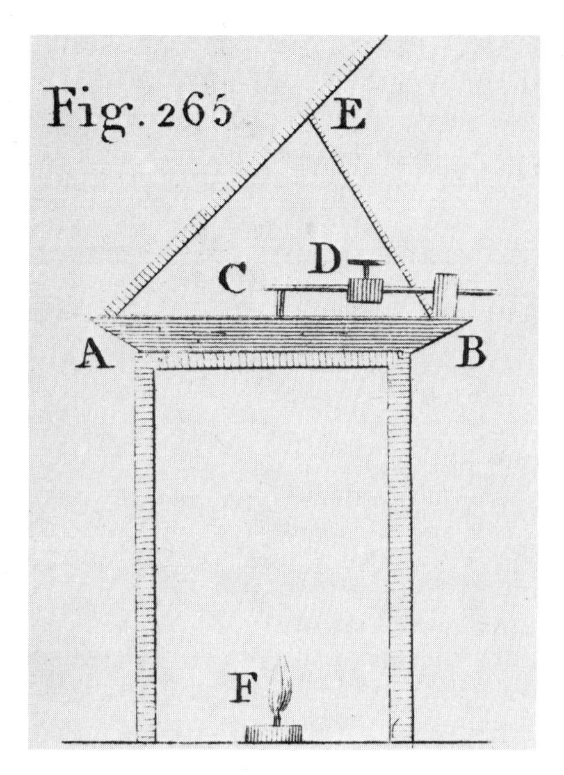

5.11 *Young's ripple tank*

time to predict the acoustic performance of concert halls, but we shall confine our attention to just two because they are fairly representative.

The first stems from experiments that were first performed as long ago as 1800. They were originally developed as a model of the behaviour of light waves by Thomas Young at the Royal Institution, and a piece of apparatus dating from that time and firmly believed to be that used by Young is still in existence. 5.11 is a drawing taken from Young's publication dated 1802. Similar ripple tanks are widely used nowadays in schools for demonstrating all kinds of features of the behaviour of waves. In order to take photographs it is convenient to arrange a bright source of light as shown in 5.12, and shadows of the ripples on the surface are then projected on to the flat surface and the camera can photograph them quite easily. In order to represent empty space for our acoustic studies we surround the tank with a shallow 'beach' of coarse foam rubber which effectively absorbs the ripples.

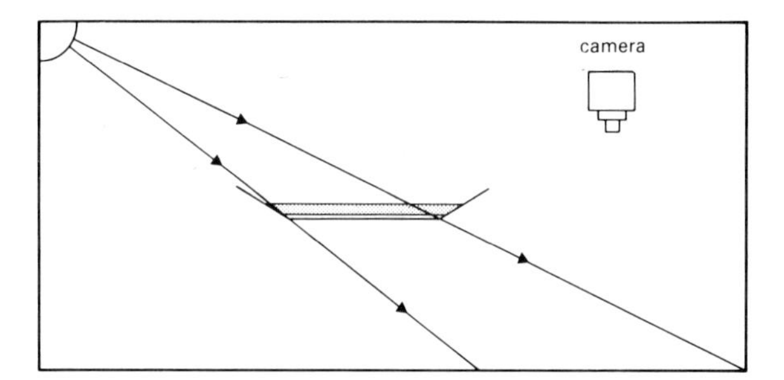

5.12 *Arrangement of ripple tank*

5.13 shows a set of photographs taken with two bars of metal in the tank to represent the ground and a single wall. The three distinct reflected waves can be clearly seen as in 5.3 at the beginning of the chapter.

These photographs are new, though the method is an old one and is probably more useful for demonstrating principles than in research. The other method to be described, however, may well have greater possibilities in research. It involves building a scale model of a projected concert hall and using actual sound waves to test its behaviour.

The problem, of course, is one of scaling. At the beginning of the discussion of Sabine's work we stressed the time element and found that the larger the hall, the longer would be the reverberation time. But a model – say at 1/8 scale – would have a volume of less than 1/500 of that of the hall, and on that account alone the time of reverberation would be much reduced. There are many other scaling factors to be considered too. For example, we saw in chapter three how a volume of air in a cavity has certain resonant frequencies which are related to the number of wavelengths that can be fitted in in a particular direction. Do we therefore need to scale the wavelength of the sound if we are to use it in a model? If so, how can we do it? Can we match the absorption coefficients of wall finishes and 'scale materials'? How can we represent scaled-down audiences?

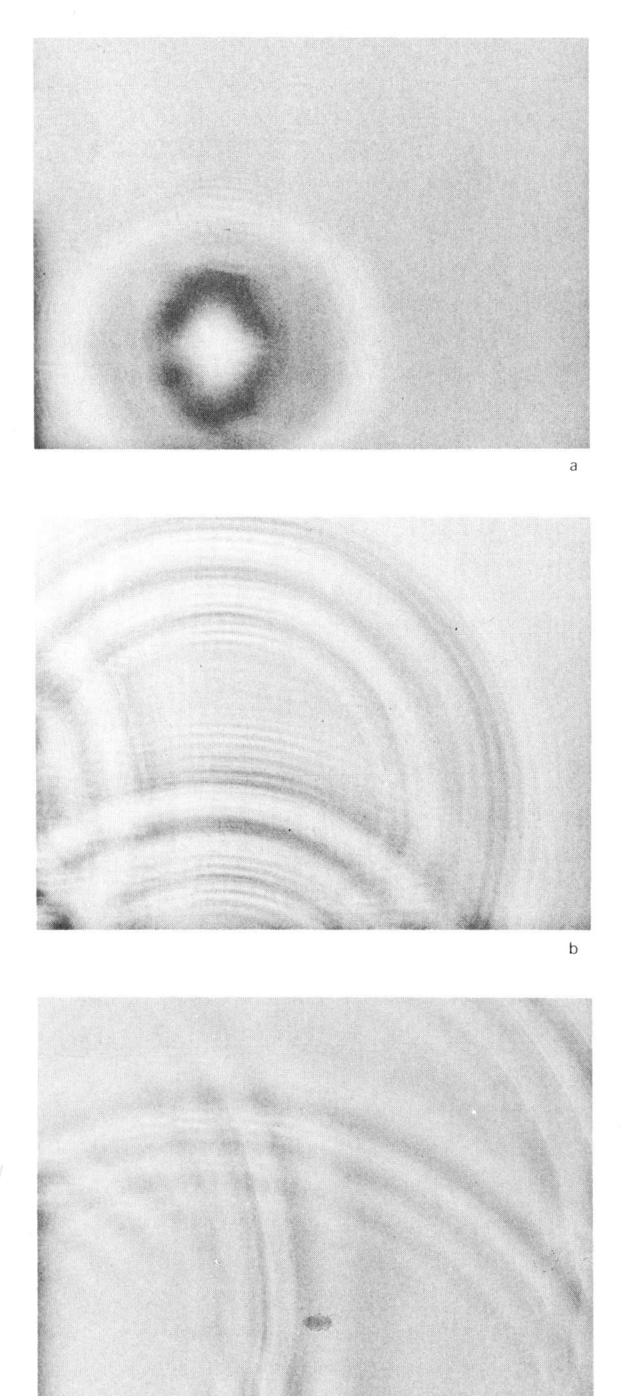

5.13 a–c *Ripples in successive positions reflected from two bars at right angles: this corresponds to the acoustic situation shown in 5.3*

5.14 *The inside of an anechoic chamber*

These and a great many other questions are being answered by a team at the BBC research centre at Kingswood Warren and it turns out that one very ingenious trick solves quite a few of the problems at once.

The trick consists in first recording test sounds in as dead an enclosure as possible. A so-called 'anechoic' chamber – a room completely devoid of echo or reflection – is used for this. 5.14 is a photograph of such a room; the wedges of foam absorb very nearly all the sound falling on them. The recording is now played back at eight times the normal speed into a loudspeaker in the model. 5.15 a shows a view of a model of the BBC Music Studio at Maida Vale and 5.15 b shows a similar view of the studio itself. The effect of playing back at eight times the speed is first of all that the frequencies go up eight times and so the wavelength of the sounds produced – since we have not changed the velocity of sound – are one-eighth normal; that is, they match the scale of the model. Secondly, the rate at which sounds are produced is eight times faster; for example, the syllables of speech follow each other eight times faster and so the time intervals again match the scale of the model. The resultant sound is then picked up by a microphone in the position of the audience and the result again recorded. Finally the tape is played back at one-eighth of normal speed so that all the frequencies, time intervals,

5.15 a and b *Photograph of (a) the model of the studio made at Kingswood Warren and (b) the actual BBC studio at Maida Vale*

reverberation times, echo delays that there may be are restored to normal scale. The resulting sound is a very close approximation to that in the real studio. But it is now very cheap and easy to change the model – to add reflectors or absorbing material, to change the angle of a flare and so on, and to be in a position to predict exactly what the effect of such a change on the real studio would be before committing oneself to extremely expensive alterations in the real hall. Various materials have been discovered which have the right scaled absorption properties – such as perforated plastic sheet which behaves like perforated acoustic tile, and polystyrene foam models can be made to behave like people (5.15 c). We have used transparent scale models – such as the one shown in 5.16 with scaled-down sound waves and the magic wand described earlier to explore sound distribution.

5.15 c *Plastic material which acts as scale acoustic tile and polystyrene foam man-equivalent for use in the model*

5.16 *A transparent acoustic model used in the author's laboratory for exploring sound fields with a 'magic wand'.*

ECHOES AND WHISPERING GALLERIES

A specific echo is one of the most irritating of all acoustic defects and there are many notorious examples of public buildings and theatres which provide this phenomenon if the listener occupies certain positions. A simple demonstration of the

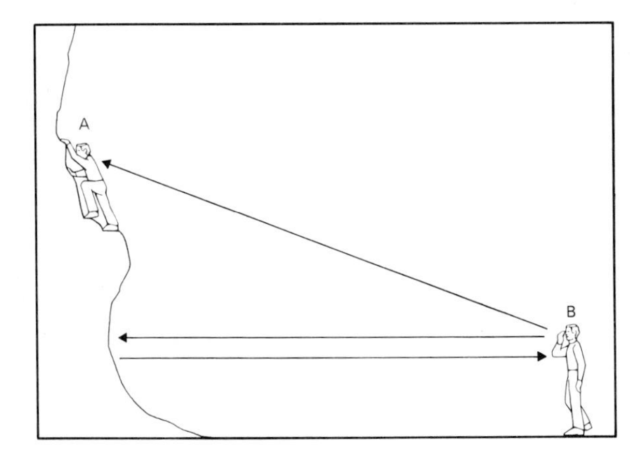

5.17 *Echoes on the mountain-side*

annoyance can be performed with any tape-recorder that has separate 'record' and 'play-back' heads. Speak into a microphone connected to the record head and, with a tape running as slowly as the machine will permit, feed the monitor output from the play-back head to a loudspeaker. The record head is always reached first by the tape, so that one can in fact use the monitor system to listen to what is actually on the tape, but of course it takes a fraction of a second for the tape to travel from the record head to the play-back head. Consequently the voice coming out of the loudspeaker is an echo of what goes into the microphone. It is very difficult to speak normally with this irritating echo repeating everything you say half a second or so later, and you will find yourself speaking more and more slowly and probably giving up the experiment.

An echo can arise simply because there is a large flat wall sufficiently far away for the sound to take an identifiable time to travel to the wall and back. What matters, of course, is the path difference between the direct and the reflected sound. For instance in 5.17 our little men A and B have gone mountain climbing. A is already up the mountain, but B is still in the valley. If B shouts to A, A will not notice an echo, as it takes almost as long for the direct sound to reach him as it does for the sound reflected from the mountain-side. B, however, will himself hear an echo with quite a long delay because he hears his own direct sound instantly and has to wait for the reflected sound to travel to the mountain and back, before he hears the echo. Echoes arising from large flat walls are rare in buildings and can easily be cured by treatment with absorbent material. Most of the notorious echoes, however, arise from the existence of curved surfaces which have a focusing effect on the waves at certain points. For example, a spherical dome which happens to have its centre of curvature at floor level can lead to very awkward echoes. 5.18 and 5.19 show successions of ripples in a crude model representing such a dome. The concentration of waves is very noticeable and the waves can travel back and forth many times before dying out, so that any sound produces very annoying multiple echoes. This kind of focusing is responsible for some of the well-known 'whispering-gallery' effects, where a speaker at one side of a curved room can be heard with astonishing clarity at another point a considerable distance away. Further examples of focusing are shown in 5.20 and 5.21. 5.20 is for a room with circular plan and demonstrates the whispering-gallery effect. 5.21 shows a dome in a theatre – no longer existent – which produced an alarming echo in parts of the balcony. 5.22 shows that domes can be used provided the dimensions are carefully chosen – here the radius is twice the height and the waves spread out uniformly and can be absorbed by carpets, etc. The cure for an echo arising from focusing is either to coat the dome surface with absorbent material or to provide absorbent and reflective obstructions to break up the wave before it reaches the dome – as, for example, has been done in the Royal Albert Hall (5.23).

VARYING THE ACOUSTIC PROPERTIES TO THE OCCASION

We have discussed some of the more obvious ways in which acoustic properties can be modified in a more or less permanent way by adding absorbent material or by introducing reflective flares. But suppose either that a hall has too small a time of reverberation and needs 'livening

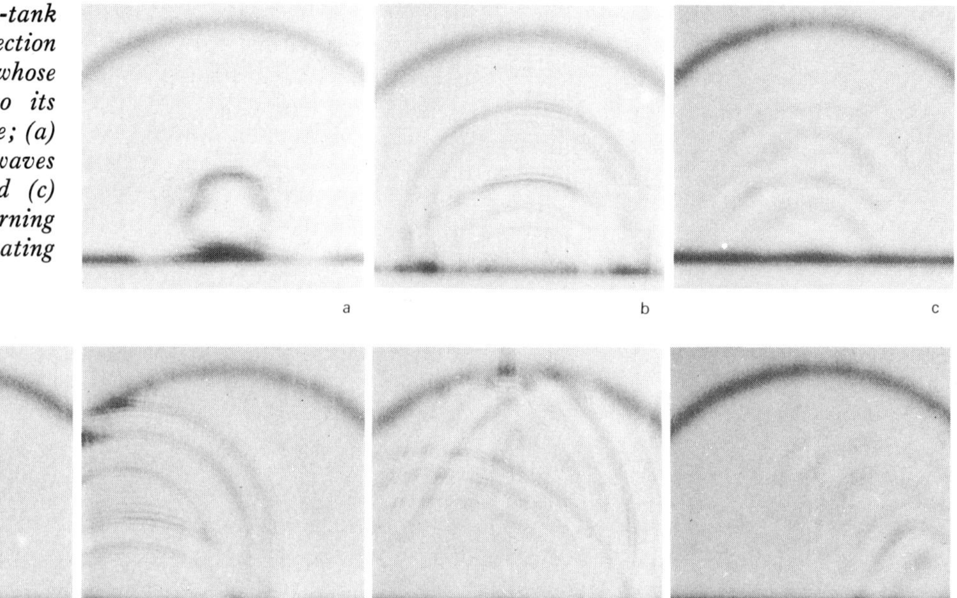

5.18 a–c *Ripple-tank series showing reflection from a dome whose height is equal to its radius of curvature; (a) and (b) show the waves spreading out and (c) shows the returning waves concentrating back on its source*

5.19 a–d *The same dome, but with the source at one side showing how strongly the waves are concentrated on a corresponding point at the other side*

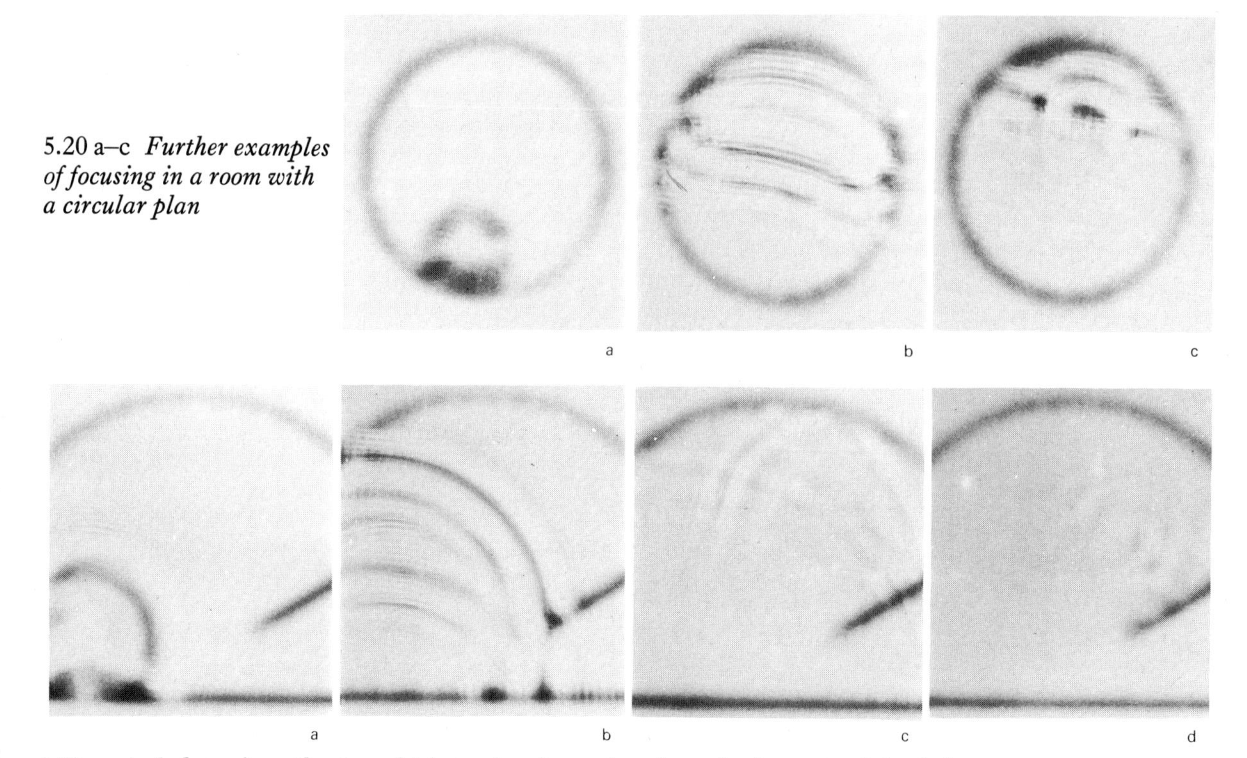

5.20 a–c *Further examples of focusing in a room with a circular plan*

5.21 a–d *A dome in a theatre which produced an alarming echo in parts of the balcony*

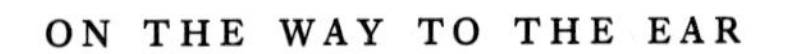

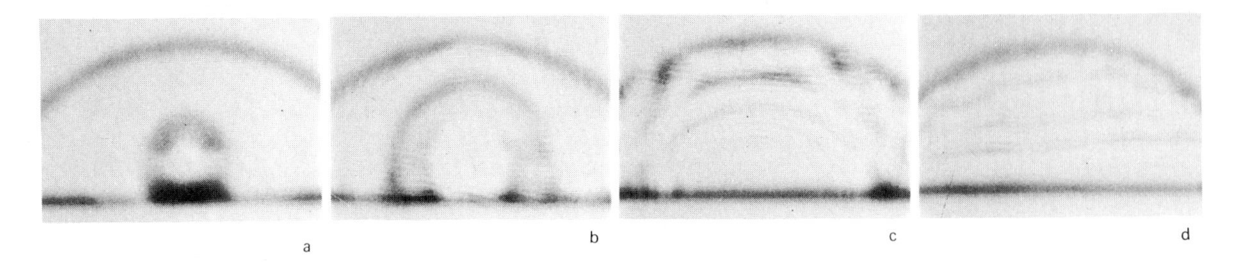

5.22 a–d *Carefully chosen dimensions. Here the radius is twice the height, and the waves spread out uniformly*

up' or that a room needs to be very dead for one purpose and lively for another – can anything be done?

Various attempts have been made from time to time to provide changeable panels with hard material on one side and absorbent material on the other. Usually, however, the areas needed to make an appreciable effect are very large and hence the changes can only be done with considerable effort or with expensive machinery. It has been something of a dream for acoustic designers that it might be possible to introduce some kind of electronic system that would permit the properties of the hall to be changed at the flick of a switch. The possibility raises a great many interesting questions about who should set the controls– the performers, the conductor, the music critics, the composer, etc. But we will shelve speculation on that problem, fascinating though it is, and consider how nearly the problem has been solved.

5.23 *Royal Albert Hall dome*

It is, of course, quite easy to change the quality of sound produced in a hall if the resultant sound is being fed elsewhere – as for example in broadcasting and recording. The addition of reverberation, for example, is widely practised. It can be done in several ways. A little of the sound being produced can be fed to a loudspeaker in a large reverberant, empty room and the resultant sound picked up by a microphone. Various fractions of this reverberant sound can then be mixed in with the original to give the desired effect. A more compact but surprisingly effective alternative is to feed part of the signal to a loudspeaker unit clamped to a metal plate mounted on springs which is fitted with a pick-up. The vibrations are present in the plate for some time and the resultant signals can be mixed in as were those from the reverberant room. Purely artificial reverberation is also possible with the help of a specially designed recorder which may use an endless loop of tape or a magnetic disc or drum. The principle is to record in the normal way and to use a large number of play-back heads at various positions so that the recorded sound is played back with varying delays. The positions of the play-back heads can be adjusted to vary the delay, and the amplitude of response from each can be controlled and mixed in any desired way. This is much the most flexible system and one can adjust the settings to represent the various reflections from the walls of a building and, given enough heads, almost any property can be reproduced.

The great problem of course is that this works splendidly on sounds being transmitted elsewhere electrically, but it is very difficult to feed the sound live back into the concert hall so that the performers and the live audience can experience it. The principal reason for this has already cropped up in chapter one in which we discussed ways of starting up oscillations. If the treated sound were fed back into the hall it would itself be picked up again by the microphone and strange feed-back situations would arise.

Some success has been achieved, however, and one or two examples of modification systems – often referred to as 'ambiophony' (surrounding sound) systems – are in existence.

One way of solving the oscillation problem is to provide microphones very close to the sources of sound, so that they do not need to be very sensitive, and to play back the modified sound through a very large number of small, low-level loudspeakers. An interesting compromise between purely electronic ambiophony and the system of using an empty reverberant room has been developed by the physics department at UMIST and applied to a large lecture theatre there which was designed primarily for speech and is therefore somewhat dead for musical purposes. A large cavity between the suspended ceiling and the structural ceiling is used and the sound from the hall is fed in non-electronically by the simple expedient of opening slots in the ceiling. The reverberant sound in the cavity is then picked up by the microphone and fed to a large number of small speakers all round the main theatre.

Modifications using a different principle have been put into effect at the Royal Festival Hall. The system used there is known as 'assisted resonance' and aimed at increasing the reverberation time at the low-frequency end of the spectrum. The designers have aimed to defeat the feed-back problem by using very carefully selected positioning of microphones and loudspeakers in relation to the modes of vibration of the air in the hall. In chapter three we saw some examples of resonance in three-dimensional vessels and of course a concert hall has an extremely complicated pattern of resonances at all frequencies. If a loudspeaker is set up at a particular point and generates sound at a particular frequency, a complex standing wave pattern will be set up in the hall and it will be possible to find places where there are pressure maxima and minima. A microphone placed at a precise maximum and feeding the loudspeaker through an amplifier will then be likely to prolong reverberation at this frequency while much less likely to interact with other loudspeakers and microphones at other positions. In practice the patterns are so interwoven that both the microphones and loudspeakers have to be tuned so that they only respond at the required frequencies. This is done either by placing them in Helmholtz resonators (see 3.22) of the appropriate size or by placing them at the end of tubes of the appropriate length. The system is complicated to set up and adjust, but has undoubtedly had a very beneficial effect both from the point of view of audiences and performer. Modes in a hall can be demonstrated by means of the scale-model principle – provided of course that the pitches of the sounds are suitably scaled up too.

The model already shown in 5.16, which represents a hypothetical hall design at 1/16 scale, has been used for this purpose. A loudspeaker is placed in the corresponding position to the source of sound and is fed with tones of various frequencies. The three-dimensional nodal pattern can then be explored using the magic wand either inserted through a universal joint in the roof or projecting up through a slot in the floor. Scaling requires the frequencies to be multiplied by 16 and hence tiny speakers are adequate.

RECORDING AND TRANSMITTING MUSIC

We have already introduced the idea that singing in the bath – or in any surroundings other than empty space – introduces another formant characteristic on top of those already implicit in the instrument or voice producing the sounds. If that sound is now picked up by a microphone and recorded or transmitted, the properties of each part of the system – microphone, amplifiers, recording systems, play-back system, loudspeakers – all impose their own formant and one begins to wonder if the resulting wave form at the end even remotely resembles the sound produced to begin with. In fact it very rarely does. But the truly astonishing thing is that the ear and brain do not seem to mind. Fortunately the formants are not usually highly complicated. More often than not they represent a cutting down of components at both high and low frequency ends and it is truly amazing what allowances can be made. In 5.24 a we see the complete wave form of the word 'sound' recorded on as faithful a system as possible (i.e. a system with all the formants having uniform level amplification over the range). In 5.24 b we see the complete wave form of the same sound after passing through a system which transmits practically no frequencies below about 700 Hz or above about 1400 Hz. In other words we are using only about one octave out of the nine or so octaves to which our ears are sensitive. The wave form looks quite different and yet the word is completely intelligible. In 5.24 c the frequencies present lie between 2800 and 5600 Hz, and although very thin and sibilant in quality, the word can still be understood. We shall have more to say about this in the next chapter.

Why then do we bother about hi-fi? Although

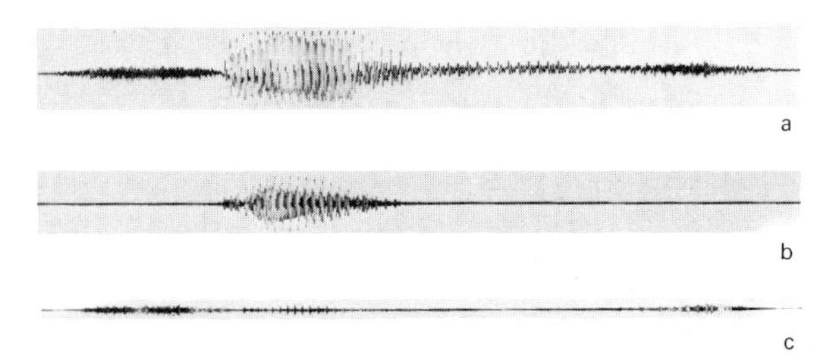

5.24 a–c *(a) The complete wave form of the word 'sound'; (b) the complete wave form of the same sound after passing through a system transmitting no frequencies below 700 Hz or above 1400 Hz. (c) Frequencies between 2800 and 5600 Hz. Although very thin in quality the word can still be understood*

the sounds from a poor system are decipherable, the ear and brain have to work hard to do the disentangling, whereas if we are presented with an undistorted version the listening is much easier and pleasanter, and also we shall be able to hear an enormous number of features of the sound which – though not absolutely essential for recognition – nevertheless add enormously to our enjoyment and are absent from narrow band signals.

On the whole, up to now efforts to produce good-quality signals have concentrated on keeping the formant at each stage as simple as possible – ideally a straight line indicating precisely equal amplification at all frequencies. Or alternatively – especially in recording –in providing formants at one stage which are complementary to formants at another so that the end result of all of them is level response. Thus the high-frequency components might be overamplified at one stage, and then after the play-back stage a reverse formant which cuts down the high frequencies can be applied. The end result is a uniform response for the signal which is desired, but the high-frequency noise which intrudes at the recording and play-back heads – i.e. after the initial treble boost – is cut back and the result is much less noisy reproduction. This is just one of many ways in which formants of various kinds can be deliberately used to improve the quality of the final result.

Recently, however, a completely new method of preserving fidelity is coming into use and it is related to our discussions on the use of computers as sources of sound in the last chapter. The system is sometimes described as 'pulse code modulation' or p.c.m. The idea is first to change the wave into a set of numbers representing the pressure in the wave at successive time intervals. This is exactly the reverse of the process of generating sounds from a computer. Once in digital form the numbers can be transmitted over a telephone line, a radio link, or recording. The result is then played back by a digital-to-analogue converter of exactly the kind we described in the section on computer music. The tremendous advantage is that there can be quite a lot of noise in the transmitting or recording process without interfering with the recognition of the *numbers* or *digits*. Thus when the signals are decoded the original quality is preserved. Of course it is possible that errors may occur in the transmission of the digits themselves. If an error occurs it introduces a very tiny click into the sound. But even the effect of this can be reduced by a very clever computer system. In effect what this does is to operate on the principle that the pressure in a sound wave is unlikely to change more than a certain amount – whatever the signal – between one pulse and the next. If any pulse comes along that is incorrect and represents a bigger change than expected, the system throws it out and repeats the last correct pulse; as a result the click disappears. Of course if the mistake comes too rapidly the system cannot cope, but the improvement over normal transmission is very great indeed. The same system has great possibilities in increasing the number of signals that can be sent along a channel at the same time as well as improving the quality. This kind of system is now being used not only for recording high-quality sound but also vision, and it seems clear that recording and transmission of signals in digital form is likely to overtake conventional systems – which in computer parlance would be described as analogue systems, since they operate with currents or voltages proportional to the sound pressure.

So much then for all the hazards that must be overcome by the sound waves on the way to the ear; we now reach the end of the journey through the air and must consider what happens once the waves impinge on the ear of the listener.

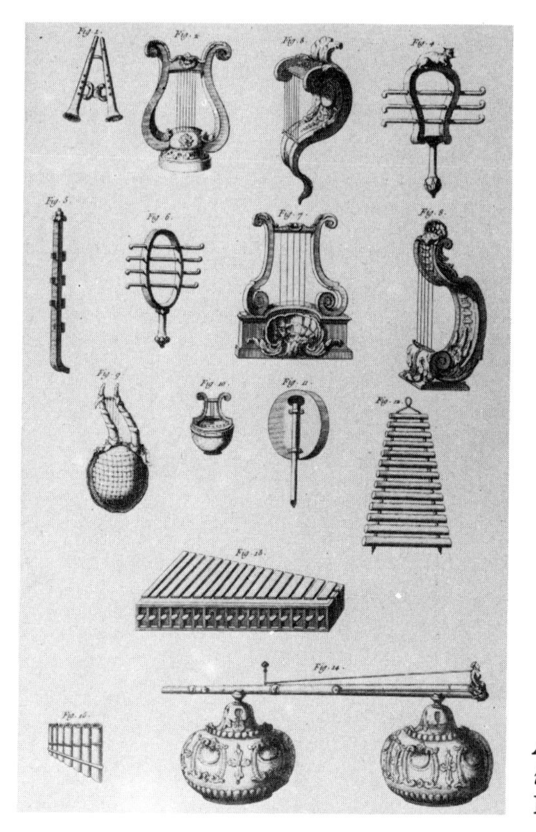

A page of musical instruments from Diderot's Encyclopédie

A woodcut by Harunobu

THE END OF THE JOURNEY

THE RECEIVING APPARATUS

Often, in simplified diagrams of the human body, or in elementary discussions of sensory mechanisms, the ear is represented as a telephone receiver which picks up sound and passes it on in some intelligible form to the brain. This picture is, I suppose, valid in the sense that the ear has a diaphragm which vibrates in sympathy with the incident sound waves, and some kind of electrical signal leaves the ear by the auditory nerve; but that is about as far as the resemblance goes. As we shall see, the ear on its own, without all its complex interactions with the brain, is a very limited organ indeed. In spite of this, however, we shall start by describing what the ear itself does, concentrating mainly on its characteristics as a piece of physical apparatus rather than venturing into anatomical or physiological discussions. To underline this point, we have deliberately omitted the well-known diagram of the anatomy of the ear that appears with monotonous regularity in textbooks; 6.1 is our substitute!

There are three obvious divisions of the ear in most mammals. First, the outer ear, which is merely a funnel to collect the disturbances in the air. It appears to have two main functions.

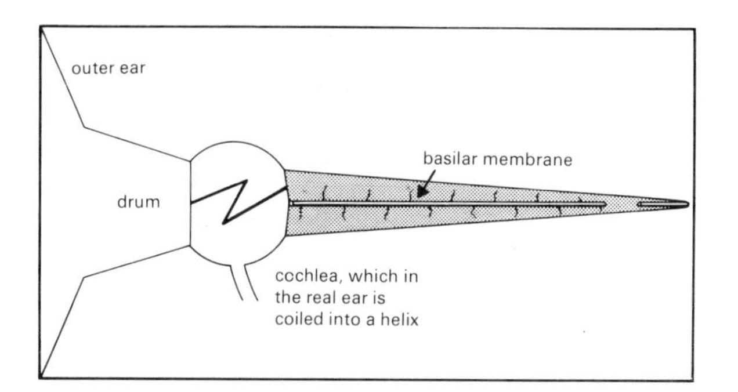

6.1 *Schematic diagram of the ear*

One is to amplify the sound simply by channelling the energy dispersed over a large area down a tube and concentrating it on to the small diaphragm or drum stretched across the end; the old-fashioned ear trumpet is an extension of this principle to increase the amplification still further. The second is to provide some directional sensitivity. This is not well developed in humans, but watch the searching rotation of the much larger outer ears of a horse or an Alsatian dog and you will see this aspect amply demonstrated. Some workers have claimed a third function, that the irregular shape of a human ear actually modifies the sound. The fact may well be true, but whether accidental or the result of evolutionary specialisation, and to what purpose, remain unanswered questions. The drum or tympanum is under involuntary control by muscles which can change its tension, but this operation seems to be a protective device to minimise damage under conditions of extreme loudness rather than an essential part of the normal hearing mechanism.

The waves in the air arrive at the outer ear and are channelled down to the drum which follows the variations in pressure more or less faithfully. The linkage of small bones in the middle ear, which is the second division, behaves like a simple lever system and reduces the relatively large amplitude of movement of the drum to a much smaller movement of another diaphragm – the oval window.

The middle ear is a closed chamber and hence variations in the total atmospheric pressure would cause distortion of the drum, so a leakage path is provided which permits the pressure in the middle ear to be equalised with that outside. A narrow tube – the Eustachian tube – links the chamber to the back of the nose and throat. Blockage of this tube, for example, as a result of infection, prevents equalisation of pressure and the resultant distortion of the drum can cause partial deafness and discomfort, as when one has a cold; similarly very sudden changes in atmospheric pressure when flying in an aeroplane, or entering a tunnel when in a train, can cause temporary effects on hearing, as it takes time for the pressures to equalise.

The third division is the inner ear, in which the vibrations of the oval window are converted into signals which can be fed to the brain along the auditory nerve. The crucial component is the cochlea – a conical tube normally wound into a snail-like spiral, but shown straight on our diagram. It is divided down the middle by a partition known as the basilar membrane and contains a fluid. The vibrations of the oval window set up vibrations in the fluid which travel along the upper half, slip through a small opening in the basilar membrane at the small end of the tube and travel back along the lower half, finally emerging from a second diaphragm back into the middle ear.

This is an over-simplification, but it is clear that a crucial factor in hearing is the interwoven pattern of waves in the fluid and on the basilar membrane itself.

The membrane has hair-like cells attached to it – 25,000 or so in man – and relative movements or vibrations of these hairs in the fluid provide the transition from the mechanical waves into nerve signals of some kind.

This is one of the points at which controversy enters and there have been many theories of the precise mechanism by which this transfer occurs. Early among these theories was the idea that each fibre responded by simple resonance to a particular frequency-component in the incoming sound. This, clearly, is inadequate as it leaves unexplained so many properties of our hearing system – not least our rapid response to transients (which was discussed in chapter two). Furthermore, physiological experiments on animals have shown directly that the signals passing along the auditory nerve to the brain still retain some characteristics of the original wave form and are not merely coded indications of frequencies.

HOW SENSITIVE IS OUR HEARING?

Before delving further into the fascinating, and still relatively obscure, problem of how our ears work, it would be useful to remind ourselves of some of the facts about our complete hearing system.

In terms of amplitude first of all, the so-called threshold of hearing, the tiniest sound that we can just hear in conditions of absolute silence, represents a change in pressure of something like 2 parts in 10,000 million of the atmospheric pressure. At the other end of the range, a sound which just ceases to be heard in the ordinary way and becomes painful corresponds to a change in the atmospheric pressure of about 2 parts in 10,000. (Or in more scientific terms, the threshold of hearing involves a pressure change of 2×10^{-5} Newtons per square metre, and the threshold of pain, 20 Newtons per square metre. Atmospheric pressure is about 100,000 Newtons per square metre.) Thus the range covered is rather more than a million to one.

These figures must be taken as very rough guides because sensitivity changes from person to person, it varies in any one person with age, it varies with the kind of sound used in the test (for example, it is quite different for pure tones of different frequencies) and also with the surroundings and the conditions under which the test is made.

A great deal of information is now available on the kinds and levels of sound that cause discomfort or even permanent disturbance of the hearing function, and noise is one of the important factors in our contemporary ecology. The measurement of noise, or of sound levels, is really a very difficult problem indeed, primarily because we are trying to find purely physical ways of measuring a response that is largely psycho-physiological. In earlier chapters we have touched briefly on the relationships between intensity (the physical quantity) and loudness (the perceived sensation) and also between frequency (another physical quantity) and pitch (the corresponding perceived sensation). Now we must delve a little deeper and first of all we should consider the general relationship that exists between stimulus and sensation for all our senses. The earliest study seems to have been by E. H. Weber early in the nineteenth century. He worked with weights and the sensation of touch. An observer was blindfolded and various weights were placed on the outstretched palm of one of his hands. It was found that when he had only very small weights already in place he was sensitive to quite a small addition. If on the other hand he was already very heavily loaded, quite a large additional weight was needed before he was conscious of a change. In very crude terms one can put over the sense of his discovery by saying that if you stick a pin in someone's hand the first thousandth of an inch is quite noticeable and painful; but if the pin is already in half an inch, another eighth of an inch or so makes little difference!

In the realms of vision and hearing Weber's observation becomes much more precise and, in scientific terms, the law that emerged from his work – the Weber–Fechner law – can be stated as 'the increase in stimulus (physical quantity) needed to produce a given increase in sensation (perceived quantity) is proportional to the existing level of the stimulus'. In other words, suppose we are listening to a sound which is fairly quiet and we then double the physical quantity – in this case the intensity of the sound – and find that we can perceive a change. If then we listen to a sound whose level is already, say, twenty times that of our first signal, we shall need to add twenty times the amount we added to the first sound if we are to experience the same *apparent* perceived increase in the sound level. As we shall see later, the law is not really obeyed rigorously by any of our sensations, but one can think of many everyday experiences which confirm the general idea. At night when light levels are very low, motor-car headlights appear very bright; during the day, when we are already being stimulated by a great deal of light scattered from the sky, the additional light of headlamps is hardly noticeable. Perhaps after all there is something in the old teacher's trick of persuading a class to be quiet by trying to hear a pin drop! The extra stimulus produced by the pin dropping is very tiny and will only be perceived by the ears and brains of the listeners if the steady level of sound already there is also very low.

Sound	Level	Intensity
Threshold of pain	120 db	— 1 watt/metre^2
Riveting steel plates about 6 feet away	110 db	— 100 milliwatts/m^2
Large symphony orchestra playing fff in a concert hall	100 db	— 10 milliwatts/m^2
A pneumatic drill at about 10 feet away	90 db	— 1 milliwatt/m^2
	80 db	— 100 microwatts/m^2
Traffic in a busy city street	70 db	— 10 microwatts/in^2
A crowded restaurant	60 db	— 1 microwatt/m^2
	50 db	— 0.1 microwatt/m^2
Background noise in a town at night	40 db	— 0.01 microwatt/m^2
	30 db	— 0.001 microwatt/m^2
	20 db	— 0.0001 microwatt/m^2
A mosquito passing one's ear	10 db	— 0.00001 microwatt/m^2
Threshold of hearing	00 db	— 0.000001 microwatt/m^2
		$= \frac{1}{1,000,000,000,000}$ watt/m^2

6.2 *A list of common sounds together with scales of ratios expressed in decibels, and sound intensity*

We have already met with this kind of law in our discussions of pitch and frequency in chapter one; 1.14 showed how, by switching to a logarithmic scale, we can represent the intervals that *seem* to be equal to the ear (the octaves, for example) by equal steps on the diagram and we then find that the frequency is being *multiplied* by two each time. Fechner claimed that this same reasoning applied to all sensations, and although this is not now generally accepted, the logarithmic nature is so much nearer the truth than linear behaviour that logarithmic scales are widely used in loudness and pitch measurements.

I promised to keep mathematics out altogether, but the term 'decibel' has become part of everyday language and must therefore be explained, but it is almost impossible to explain it without any mathematics at all. The first point to get clear is that the decibel is not, in itself, a measure of loudness but is concerned with the ratio of two quantities. Secondly, although measured on physical apparatus, it is on a logarithmic scale and so corresponds very roughly to our response to the quantities being measured.

In 6.2 we have shown on a vertical scale on the extreme left a number of different common sounds. They have been arranged on a scale which corresponds to our sensation of hearing – not to the measured sound pressures or intensities. The middle scale is of ratios expressed in decibels and one can see that each of the big divisions representing 10 decibels is equal in length. But now look at the right-hand scales which show the corresponding sound intensity – the rate at which sound energy is falling on the ear measured in watts per square metre. These scales are clearly not linear and one can see that each increase of 10 decibels corresponds to *multiplying* the intensity by 10. (The mathematical relationship is simply that the number of decibels is 10 times the logarithm of the ratio being considered; thus if we *double* the intensity, we take the log of 2, which is 0.3010, and multiply by 10 and obtain approximately 3 decibels or 3 db; similarly multiplying by 10 corresponds to 10 db, by 100 to 20 db, by 1000 to 30 db, and so on.)

Although this argument all sounds very neat and convincing, it is still, I regret to say, an over-simplification. It would be out of place in the text of a book of this type to take the argument very much further, but it is important to realise that our hearing sensitivity does vary enormously with frequency, with the kind of noise being used, and in particular we can get into very deep water indeed if we try to make comparative measurements with a standard sound that is very different from the sound being measured.

If we compare the sound being measured with a standard sound of exactly the same type then we can quote the 'sensation level' in decibels. If, however, we compare our unknown sound with a pure tone of 1000 Hz and adjust them so that they seem to be equally loud then the ratio of the standard sound to the threshold of the standard sound, expressed in decibels, is called the loudness in 'phons'.

But let us leave these complications and consider some of the interesting consequences of the Weber-Fechner law in hearing. The phenomenon of 'masking' is very well known – though almost certainly not by that name – to most students. If you try to do homework, or to study, in a quiet room, then every time the dog next door barks, or the tea cups rattle in the kitchen, or someone in the next room talks in a loud voice, you are distracted. Turn the radio on to some rather dull and relatively uninteresting background music, however, and the distractions cease. You are using the masking effect predicted by the Weber–Fechner law, and by creating an existing level of sound, the extra sound level necessary to impinge on your consciousness is much greater than if the room were quiet to begin with.

This is a straightforward masking effect. But one can play additional tricks here that turn out to be very useful in acoustical research. The masking effect is frequency sensitive. Suppose you listen to a steady tone on the note A on which the orchestra tunes (440 Hz) at quite a high level. Now listen simultaneously to another intermittent note, say, two octaves higher (1760 Hz). You will find that you can hear the high note coming in even when very quiet indeed. On the other hand the increase in level of the 440 Hz note needed before you notice the difference is very high. In this case the Weber–Fechner law seems to work only when the stimuli are of the same type. Why, then, does the radio help concentration? Simply because the background music contains stimuli at most of the frequencies that matter. In fact white noise would be very much better at masking out unwanted sounds. A fascinating development of this has been used by Professor Schouten and others in acoustic research. If noise which has been filtered so that it contains only a certain range of frequencies is used, then it will mask only those sounds whose frequencies lie in the same range. Thus if we are listening to a mixture of sounds and wish to focus our attention on one sound rather than others we can eliminate the unwanted sounds by masking with filtered noise and in this way can actually determine the frequency region in which the sounds being studied exist. We shall come back to this trick later as it plays an important part in trying to understand the way in which we hear mixtures of tones.

Before we finally leave the question of loudness let us consider just what happens when we try to make measurements. If we take a large number of observers and a large number of pairs of pure tones of different frequencies and ask the observers to adjust pairs of tones so that – although different in pitch – they nevertheless seem to be equally loud when heard alternately, we eventually arrive at a series of curves which are usually called the Fletcher–Munson equal-loudness curves; they are reproduced in 6.3. If you first concentrate on the vertical line at a frequency of 1000 Hz you will see that the curves are labelled 0, 20, 40, 60, etc., db. These are the decibel equivalents of the ratios of the intensity levels of a series of 1000 Hz tones with respect to the threshold of audibility at 1000 Hz – the weakest tone at that frequency that you can just hear. We can therefore refer to these as loudness levels in phons. Now concentrate on any one of the thick curves running through these points – say that at 60 db. A subject who is asked to compare the loudness of a sound of any frequency with a 1000-Hz tone at 60 db above the threshold would deem them equal if the intensity level of the sound being measured were that given by the corresponding point on the line. All tones on this line would therefore have an equivalent loudness of 60 phons. Notice that, for very loud sounds, the variation between the physically measured intensity level and the perceived equivalent loudness in phons measured this way is not very great. At very low levels of intensity, however, the differences are considerable. This is of

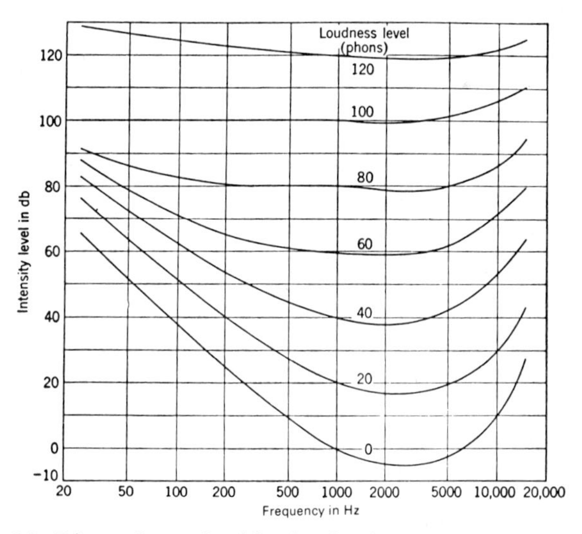

6.3 *Lines of equal subjective loudness*

course the reason why hi-fi amplifiers often have a 'loudness' control which increases the amplification at the bass and treble ends relative to the middle when the overall volume is rather quiet. This then gives approximately the same equivalent loudness over the range rather than uniform electronic amplification. If a loudness control is not fitted the same result can be obtained by turning up the bass and treble controls.

This kind of measurement, which depends on matching sounds that are intended to be *equal* in loudness, still uses *physical* methods of measuring when we want to go from one level to another. The loudness along the curve labelled 60 db is 60 phons and that of the next one up is 80 phons; the ratio of intensity levels at 1000 Hz is thus 100:1 because we can measure that on a meter. But is the 80-phon sound subjectively a hundred times louder than the 60-phon sound? The answer is no. A large series of tests have been done to try to establish a means of *subjective* loudness measurement. Various ideas have been tried. An observer might be asked to match the loudness of a certain sound heard with one ear with that of a quieter sound heard by both ears, and one might reasonably say that if the two sensations seem to be the same then one of the sounds is twice as loud as the other.

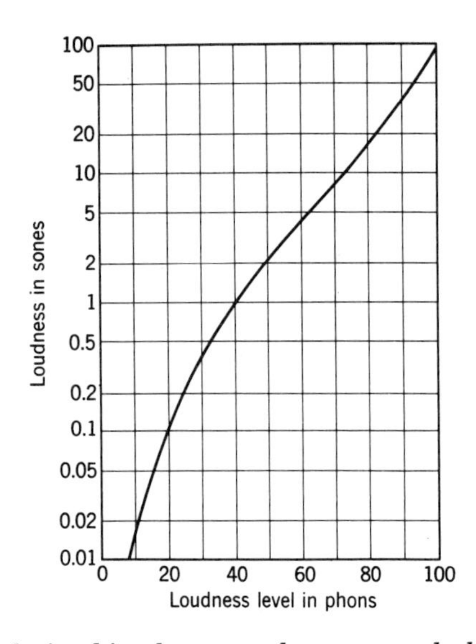

6.4 *Relationship between the measured loudness level in phons and the subjective loudness level in sones*

People have been asked just to make a subjective judgement of when one sound is twice as loud as another. Other researchers use tones of quite different frequencies and, after first setting both tones to equal loudness, use the two together as a reasonable approximation to a sound that is twice as loud as either. As a result of this kind of experiment the curve of 6.4 has been produced; it is of course a composite of many hundreds of measurements with many different observers. The horizontal scale is the equivalent-loudness level measured in phons by the technique of matching equal sounds, and the vertical scale is the *subjective* loudness measured by the methods just described in a new unit called a sone.

Quite arbitrarily a sound of equivalent loudness level of 40 phons is defined as 1 sone – it is a fairly quiet sound and yet easily heard in a quiet room. A sound of subjective loudness 2 sones would then seem to be twice as loud subjectively but would turn out – from the curve – to have an equivalent loudness of 49 phons. For loud sounds it turns out that for every 9 phons the sound seems to double in subjective loudness measured in sones. For weak sounds, however, the subjective loudness increases much more steeply.

This may well appear to be a highly confusing topic and I do not propose to follow it any further, but I hope that at least two points have emerged – one that subjective measurements are not easy to make, and the second that it nevertheless is possible to make some kind of sense out of the apparently complex field and that the results are surprisingly consistent when one checks one experiment against another.

As far as the mechanisms within the nerve-brain system are concerned it has been suggested that the subjective loudness of a sound in sones is connected with the number of nerve impulses that are transmitted per second to the brain. This would certainly tie up with the idea of measurement using comparisons of loudness with one or both ears; it would make sense that twice as many impulses would be produced in a given time if the same sound is heard by both ears as if it were only heard by one. But this is dangerous ground and we shall not speculate further!

So much for loudness. How sensitive are we to frequency differences? Again we find that the same kind of law holds, at least over part of the range, and the smallest detectable change measured in Hertz increases as the frequency of the note increases. Musically – or perceptually – the interval expressed as a fraction of an octave stays the same for intermediate and higher frequencies. Below about 500 Hz, however, the relationship is more complicated and it seems that the actual number of Hertz that can be detected as a change stays constant. Thus the *musical* interval that can *just* be detected is much larger at the low end of the musical scale.

This, no doubt, partly accounts for the much greater difficulties experienced in trying to play in tune by musicians using high-pitched instruments as opposed to those playing, say, the double bass. The complications in this area come thick and fast. We have tacitly assumed that equal intervals musically speaking really do correspond to equal ratios in frequency, and indeed, over the middle range, this is true. At the top and bottom, however, there are departures from this rule. At the high end of the range, pitch is usually judged to be lower than the simple rule would predict, and at the bottom end it is judged high. Thus, for example, at the top end of a piano an octave may require more than double the frequency, whereas down at the bottom a smaller ratio than 2:1 may be recognised as an octave.

Just as we produced a subjective loudness curve, it turns out that we can produce a subjective pitch curve, though of course there is no easy way of doing this, as one cannot imagine listening to a sound with two ears as giving 'twice the pitch' of a sound heard by one! In fact at first the idea that the frequency–pitch relationship is anything but linear seems to be quite crazy. What about our nice neat octaves which go up in equal steps and repeat with precise doubling of frequency? The answer is that these are produced essentially by comparison, and when two notes an octave apart are sounded *together* the physical shape of the combined wave is quite significantly different when the ratio is exactly 2 from that when it differs a little from 2. (6.11 f and g from page 154 illustrate this point.)

If we now construct a relationship in a different way, sounding the notes *separately* and asking the observer to judge when one note is twice as high in pitch as another, or when one note is half-way in pitch between two others, but always using *pure tones* and never letting the subject hear the tones simultaneously, we end up with a relationship like 6.5. The subjective pitch is measured in mels. In musical terms one might describe this as melodic measurement; the more usual method would be harmonic.

What we are really saying, I suppose, is that the result obtained in psycho-acoustic experiments depends very much on the exact way in which the measurement is done. This probably accounts for the widespread controversies that occur. It is so difficult to specify precisely *all* the conditions under which a measurement is made – especially those concerning the human observers – that we perhaps should not be surprised if different investigators appear to find different results.

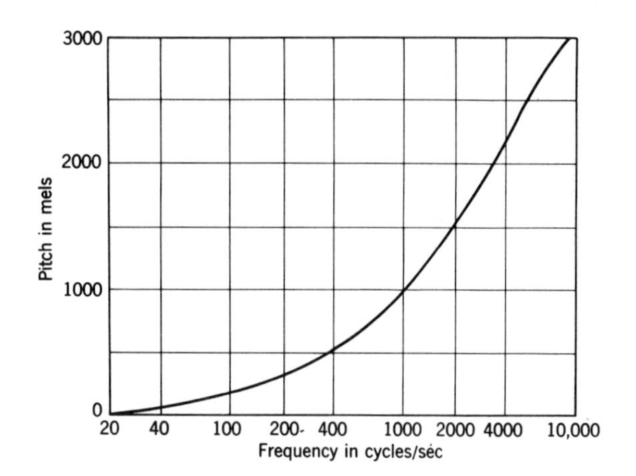

6.5 *Relationship between subjectively estimated pitch changes in mels and measured frequencies*

From the standpoint of the mechanism of hearing it seems at least reasonable to speculate that our familiar logarithmic pitch–frequency scale is tied up with the ability of the ear to recognise the shapes of waves in time, whereas the mel scale is concerned with the different locations along the basilar membrane in the inner ear at which the frequency is being interpreted.

The problems become even more fascinating and complex when one tries to make measurements with tones of very very short duration, though these are of course very important in identifying the starting transients of a particular instrument. The reader will perhaps begin to despair at this point because it appears that all the neat and simple ideas with which we began are all proving to be unsound. This phenomenon occurs in all branches of science and should not really surprise us. Life would be very dull if the simplest theories were able to explain everything precisely and many of the really fascinating ideas in science stem from the observation of small departures from simple behaviour. In the realms of psycho-physics, however, the phenomenon is far more obtrusive. Every time an observer takes part in an experiment he is changed by the experiment. *Before* he is a person who has not yet done the experiment; *after* he is a person who *has* done it and therefore now has experiences and memories available for comparison, and this may well alter his reactions the next time he performs the experiment.

But the complications and complexities of this area of study are not only scientifically fascinating but almost certainly account for the extraordinary aesthetic pleasure that music can afford. If the simple rules held precisely we should probably already live in a world in which all the possible music had already been composed and heard; what a dull world that would be!

I hope, with these words of encouragement, you can now face yet more complications. Pitch and loudness have so far been kept separate, but in fact they are interrelated. Fortunately this effect is only really strong for relatively pure tones – otherwise we should experience extraordinary key-shifts as an orchestra changed from *ff* to *pp*!

Most authors claim that if a pure tone of about 1000 Hz is suddenly increased in loudness its apparent pitch will hardly change; notes of lower frequency will appear to go lower in pitch as the loudness is increased and notes above 1000 Hz will apparently go up in pitch. My experience in demonstrating this phenomenon with a wide range of audiences of all sizes and types is that the effect varies from person to person. For example, if I play a pure sine wave at 440 Hz at a very low level, suddenly increase the intensity without changing the frequency and then take a vote on whether the perceived pitch stayed the same, went up or went down, I almost invariably find significant numbers of the audience in each category.

One final word before we leave the properties of our hearing system; as with all our senses it is much more responsive to *change* than to steady stimuli. Indeed if a subject is held absolutely stationary with his head clamped in a fixed position and a pure tone of absolutely constant frequency and constant intensity is introduced into the room, some very strange effects occur. The subject seems to become oblivious of the sound after a while or, in some cases, he may imagine changes to be taking place in the sound which have no physical counterpart. If the sound is changed, however, either by a small increment of frequency or intensity or by a momentary interruption, the subject's attention immediately returns.

It has been suggested that this may be one of the contributory factors in making *vibrato* a pleasant and acceptable feature of instrumental playing. The recurring slight changes in frequency or intensity, or both, may make it easier for the listener to hear the sound without complications. This cannot, of course, be the complete story, otherwise all vibrato would be equally acceptable. It is an all-too-familiar fact that the wrong frequency or amplitude can be aesthetically disastrous. Another interesting theory that has been suggested is that we recognise a sound which has vibrato superimposed on it as having some connection with a human performer – or possibly even with the human voice. It is certainly true that even the most accomplished singer cannot sing entirely without vibrato. It seems that the 'feed-back' mechanism, which enables us to maintain constant the pitch of the sound we are producing at a given moment, involves listening to ourselves. There is, however, a definite small interval of time which must elapse between our hearing system detecting that a drift of pitch is taking place and the correcting action being applied by the appropriate muscles. This turns out to be about 0.14 second. There is thus likely to be a wobble at about this rate which is inherent in the control mechanism and it is a curious fact that the corresponding frequency – about 7 Hz – is somewhere about the optimum for pleasant vibrato!

But we are already straying into a later section and speculating not on the physics or even physiology of hearing but on the remarkably economical way in which our brains fit clues together from all kinds of sources to give us our various sensations even when, on the face of it, we are presented with far too little information. Before touching on that absorbing topic we should spend a little time considering a question that has fascinated mathematicians and scientists ever since Pythagoras. Why do certain combinations of notes, either in sequence or together, sound pleasant and satisfying, whereas others are unpleasant or disturbing?

HARMONY, DISCORD AND MUSICAL SCALES

Gerard Hoffnung has a superb cartoon illustrating his understanding of the term 'discord' and it is reproduced in 6.6. Why do some combinations of tones sound so disturbing and why is it that, apparently, audiences can now tolerate combinations of notes that a hundred years ago would have been rejected as barbaric? The Pythagoreans have contended that the human mind prefers neat numerical relationships, and it is certainly true that two notes whose frequency ratio is, let us say, 2:3 sound pleasant together. They represent the musical interval of a fifth – C to G, for example. On the other hand, if the ratio is 200:313 – which has no recognised musical designation in the West – the result is profoundly disturbing and unpleasant.

It may satisfy mathematicians to feel that the brain appreciates the simplicity of the ratios, but a physicist would feel that the numbers themselves merely represent a language with which phenomena may be described and can have no significance in themselves. He would therefore look a little more closely at possible physical consequences of these ratios.

6.6 *Gerard Hoffnung's representation of a discord*

Helmholtz did some classic investigations just about a hundred years ago which led him to believe that the phenomenon of 'beats' was largely responsible. The effect is well known; the simplest way to demonstrate it is to ask two people to play the same note on each of two treble or descant recorders. A convenient method would be to cover the thumb hole and the three left-hand finger holes; now one of the two players should place the first finger of his right hand close to, but not firmly on, its hole so that his note is flattened slightly. Immediately the combined sound throbs, and the rate at which it rises and falls in loudness increases as the note is flattened still more. 6.7 shows an oscillograph trace of two notes played simultaneously, one of which remains at exactly the same

6.7 *Combined trace of two notes, one of which is held at a constant frequency and the other of which starts about 10 per cent lower in frequency and glides slowly upwards through the central region, where they are almost in tune, and then off to the right until eventually it is about 10 per cent higher in frequency*

frequency. The other starts at a frequency 10 per cent below, gradually rises until in the middle of the diagram it is almost the same as the first note, and then continues rising until it is about 10 per cent higher in frequency. The beat effect can be clearly seen. Helmholtz suggested that the 'beating' effect gave a rough quality to the note which was unpleasant. This might explain the rough effect of mistuning two notes that are supposed to be of exactly the same pitch. But what about mistuning of intervals such as fifths, fourths, etc.? Helmholtz argued that beats between upper partials in real instruments – which always contain a complex set of harmonics – could account for this. For example, in Table 6.I we show the sequence of harmonics of four notes forming a common chord.

Clearly there are many instances in which upper partials are identical in frequency and

TABLE 6.I

Doh	240	480		720		960		1200	1440		1680		1920		2160	2400	
Me	300		600		900			1200		1500		1800		2100		2400	
Soh	360			720			1080		1440			1800			2160		2520
Doh		480				960			1440				1920			2400	

slight mistuning would lead to beat effects amongst the upper partials. On the basis of this idea Helmholtz calculated the probable 'roughness' of intervals based on a hypothetical experiment in which two violins are imagined to start playing precisely in tune. One holds the note and the other glides slowly up through an octave. His calculated curve is shown in 6.8. The vertical scale represents the degree of roughness or unpleasantness on a scale of ten units, the highest being the most unpleasant, and the horizontal scale represents the frequency ratio, the lower note being taken as one. The interesting point emerges that if one performs the experiment using pure tones – i.e. tones with no upper partials – the result follows quite closely to Helmholtz's calculated curve! The crosses on 6.8 are experimental points found as a result of a large series of experiments on different audiences in which a series of pairs of *pure tones* were played in random sequence. They were asked to associate a grade from 1 to 10 which would be a measure of the roughness or harshness of the sound. They match Helmholtz's curve remarkably well but cannot be explained in terms of upper partials because there are none! (Or, at least, there *should* be none according to any simple theory.)

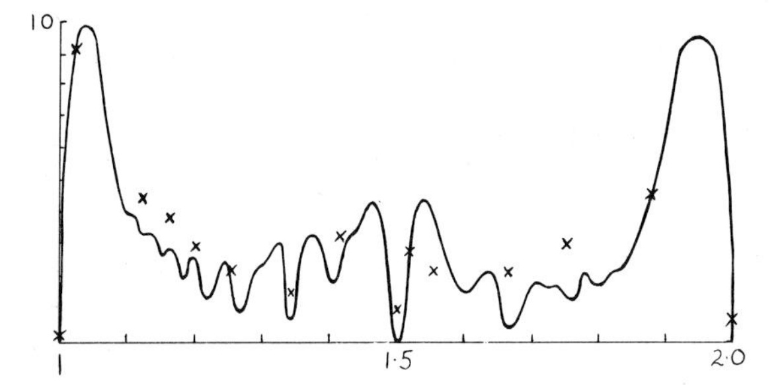

6.8 *Curves representing the degree of dissonance between a note of fixed frequency and a second note which moves smoothly from the same frequency to one an octave higher: the frequency ratio is indicated on the base line. The smooth line was calculated by Helmholtz and the crosses represent experimentally derived points from experiments with many observers*

The next point to consider is that of combination tones. It is found experimentally that when two loud pure tones are sounded together, a great many other tones are heard as well. One of the easiest to hear is the so-called difference tone. If two pure tones an octave apart are sounded very loudly and then the lower one is allowed to glide slowly up to meet the other, a strong tone which *descends* in pitch is heard. Its frequency corresponds to the difference between the frequencies of the two pure tones and, since this difference decreases, its pitch gradually descends.

These tones are generally held to have their origin in non-linear effects in the ear – probably in the middle-ear linkage. What do we mean by non-linear effects and how can they lead to the perception of tones corresponding to frequencies that are not physically present? A full explanation demands quite a bit of mathematics, but another of our mental experiments may help at least to show that these effects are possible. 6.9 shows a special pendulum that behaves in a non-linear way. ABC is the first pendulum, which is just a metal rod swinging freely at A and with a pin sticking forwards from the rod at B. DE is a lever pivoted at D and resting on the pin at B. As the pendulum ABC swings back and forth the disc E rises and falls. Now the pendulum swings symmetrically back and forth and goes as far to one side as the other. The disc E, however, only moves upwards from the point at which it rests when ABC is not swinging. Furthermore it rises from the rest point to its maximum *twice* in every *one* cycle of ABC. 6.10a shows how the two motions would look plotted on a graph with the same time scale. Suppose now instead of letting ABC

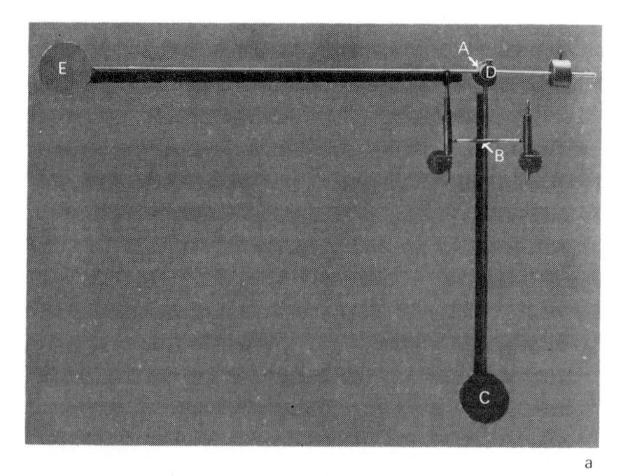

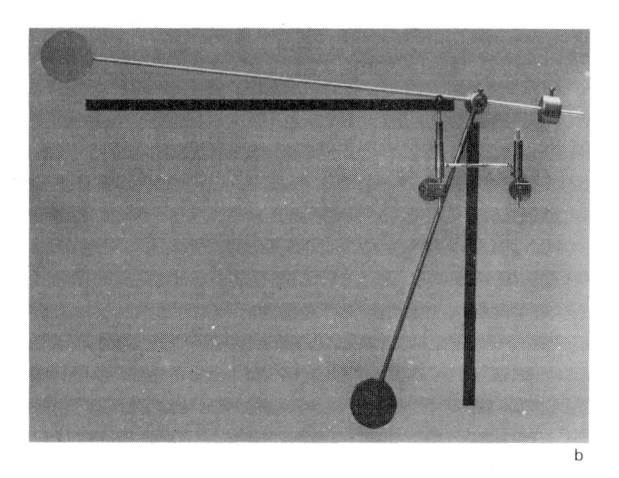

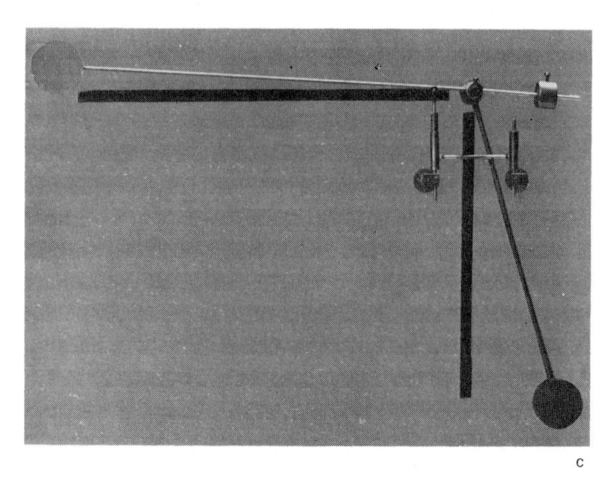

6.9 a–c *A demonstration of non-linearity: as the vertical pendulum swings through one cycle the horizontal curve rises twice above the horizontal line and never moves below it*

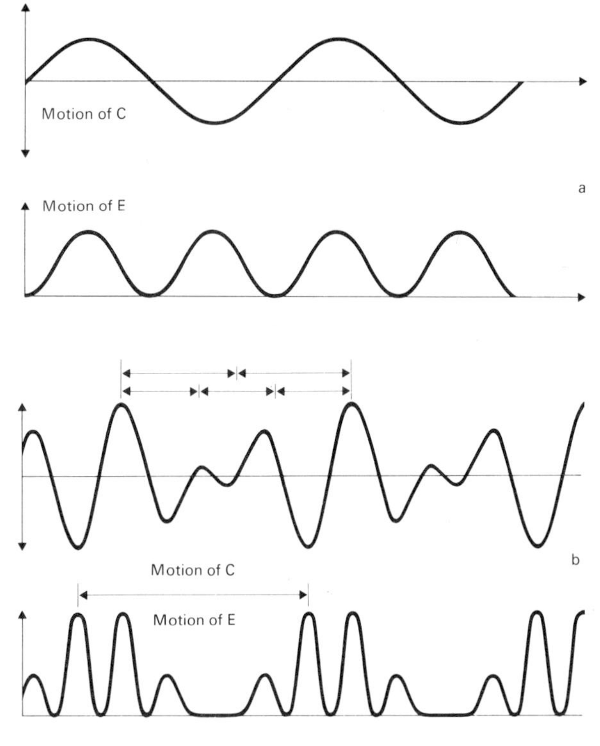

6.10 a and b *Graphs of the resultant movement of the horizontal pendulum of 6.9 when (a) the vertical pendulum swings normally and (b) when it is forced to move as though driven by two sine waves of frequency ratio 3:2*

swing freely we make it move according to a curve representing the sum of two waves of different frequencies – in this example with a 3:2 frequency ratio. 6.10 b shows how the disc would move now. The primary movement of the pendulum still is symmetrical and averages out to no displacement. But the movement of the disc is unsymmetrical and on an average has a component which corresponds to the difference of the two.

This is a very incomplete and indeed inexact demonstration, but should indicate that nonlinearities lead to more tones. Most theories suggest that they should only occur with loud sounds, however, and yet experience tells us that a pair of pure tones may be just as discordant when sounded quietly. We must therefore go on looking still further for explanations of the idea

6.11 a–g *Traces of pairs of notes in which one remains at a fixed frequency and the other is moved upwards in the ratios and actual frequencies indicated.*

(a) *1:1* *60 Hz + 60 Hz*
(b) *15:16* *60 Hz + 64 Hz*
(c) *4:5* *60 Hz + 75 Hz*
(d) *2:3* *60 Hz + 90 Hz*
(e) *20:31* *60 Hz + 93 Hz*
(f) *30:59* *60 Hz + 118 Hz*
(g) *1:2* *60 Hz + 120 Hz*

of consonance and dissonance. The mechanism is still not properly understood, but it seems increasingly clear that the mistake most of the earlier workers made was to think only about the frequencies present rather than about the shape of the wave. This is not really surprising because it was quite a long time before devices – such as the cathode-ray oscilloscope – were available to permit researchers to look at the combined wave shapes, and to calculate them in pre-digital-computer days was extremely tedious.

Consider the traces of 6.11. These are for pairs of pure tones of equal amplitude corresponding roughly to points on the Helmholtz diagram (6.9). The lower tone is kept fixed and the upper one is raised successively so that the ratios are (a) 1:1 (unison), (b) 15:16 (semitone), (c) 4:5 (major third), (d) 2:3 (perfect fifth), (e) 20:31 (badly mistuned fifth), (f) 30:59 (badly mistuned octave), (g) 1:2 (octave). 6.12 shows the Helmholtz curve again and the vertical lines correspond to the ratios of 6.11. One can see

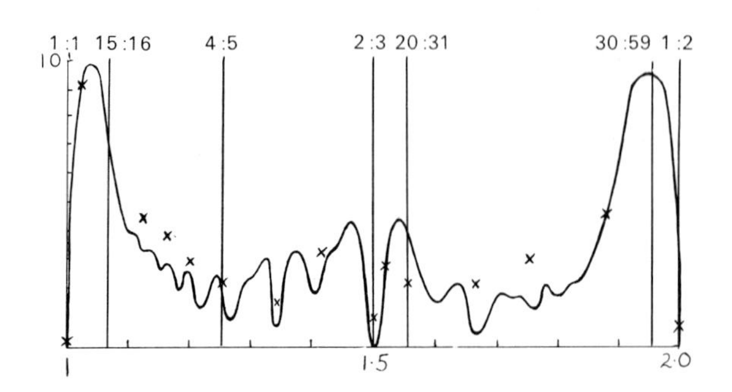

6.12 *Repeat of 6.8 with lines corresponding to the ratios in 6.11*

immediately, without delving deeply into the mechanism, that the harsher more dissonant intervals correspond to those in which changes of pressure in a given time are more complicated. If you think of our earlier discussion based on the non-linear pendulum you will realise that these complicated patterns are likely to lead to more apparent components than for the simpler ones.

But now we must return to the problem that you can hear additional components which ought not to be there even when no non-linearity is present and the sounds are very quiet. Indeed if one performs a frequency analysis of the sound of a bassoon, for example, it soon becomes clear that, even when played very quietly, there is hardly any energy at all in the fundamental frequency; it is concentrated in the third, fourth, fifth and higher harmonics. And yet one hears the fundamental note beyond

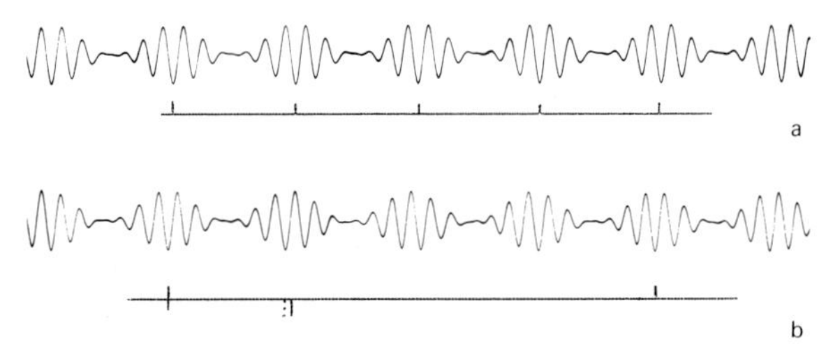

6.13 *(a) Trace of the sum of three sine waves of frequency ratios 20:24:28; vertical lines indicate repetition period. (b) Trace of the sum of three sine waves of frequency ratios 21:25:29; vertical lines* above *horizontal indicate exact repetition period, the lines below indicate repetition period on (a), the dotted line indicates near repetition period*

doubt. Look at 6.13 a which represents the sum of three components with frequency ratios 20:24:28. In other words they would correspond to the fifth, sixth and seventh harmonics of a fundamental of frequency 4. There is no component at the fundamental and yet it is clear that the resultant pattern of the wave repeats at intervals corresponding to the fundamental frequency of 4. One of the strongly held views suggests that the brain recognises this periodicity in the wave form and assumes that it stems from a component of that frequency; hence we hear it. This low component is sometimes called a 'tonal residue'.

A fascinating experiment which shows clearly that one is not merely dealing with difference tones consists in sounding three notes with ratios such as might be represented in 6.13 a and then raising each one by the same increment in *frequency* so that the differences remain the same. The difference tone should stay the same – but what actually happens is that the low note or residue *goes up*. 6.13 b shows what happens if we step up the ratios of 6.13 a to 21:25:29. There is now no precise repeat of pattern at the frequency of 4, although the frequency difference is still 4. Notice, however, that if you concentrate on two successive groups of waves there is something quite close to repetition at a frequency which is a little more than 4; arrows on both figures indicate the two repeat distances to be compared. To a first approximation this increase in repeat frequency corresponds to the rise in pitch of the residue tone which is actually observed. Does the brain then recognise a group of waves which, though not exactly the same, are still roughly similar?

A beautiful demonstration of many of these points – and of the masking effect mentioned earlier in the chapter – has been produced by Professor Schouten and it has been recorded.[1] A simple tune – the chime sequence of Big Ben – is played with each note sounded twice, the first as a pure tone and the second as a group of high harmonics giving a residue tone of

[1] 'Acoustic Demonstration by the Institute for Perception, Eindhoven', which appeared as a supplement to the *Philips Technical Review*, Vol. 24, p. 341, 1962.

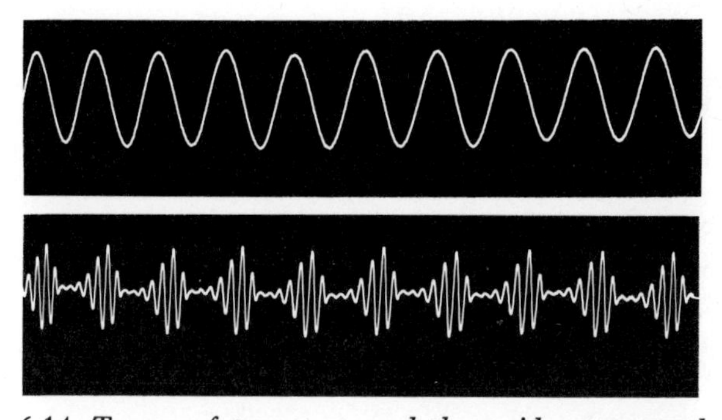

6.14 *Traces of pure tone and the residue tone used in Professor Schouten's demonstration: both sound to have the same pitch*

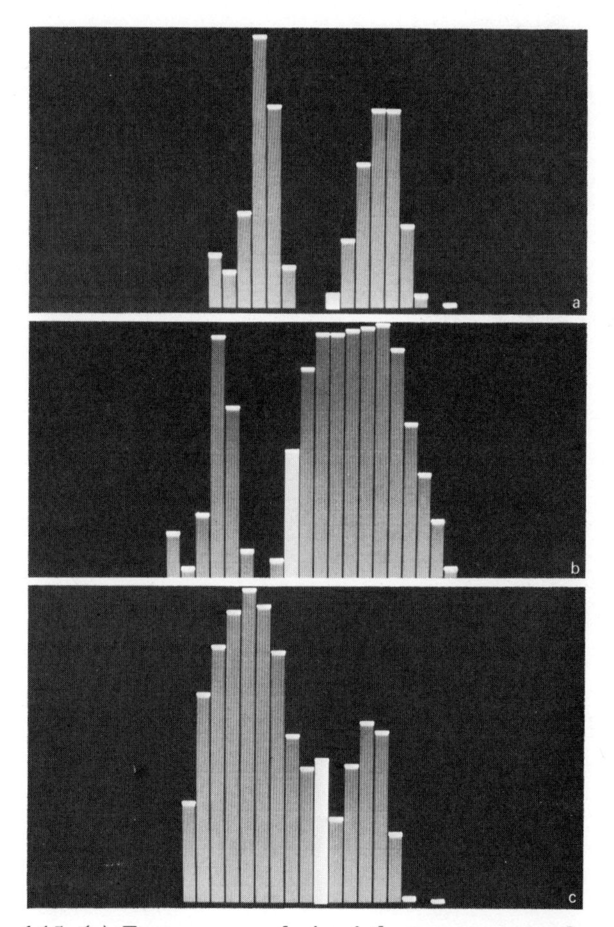

6.15 *(a) Frequency analysis of the pure tone on the left and the residue on the right. (b) Frequency analysis with low-frequency noise superimposed: only the residue tone can be heard. (c) Frequency analysis with high-frequency noise superimposed: only the pure tone can be heard*

the same pitch as the pure tone. 6.14 shows the shape of the waves for the two types. The effect to the observer sounds as though two notes of the same pitch are being used, but the second is rather harsh and 'reedy'. In the masking experiment first high-frequency noise and then low-frequency noise is mixed in. When the high-frequency noise is present only the true pure tone is heard over the top; when the low-frequency noise is present only the reedy tone can be heard. 6.15 shows at (a) the frequency analysis of the two notes, at (b) the result of adding the high-frequency wave, and at (c) the result of adding the low-frequency background.

We are now in a world of speculation in which much still remains to be done before the mechanisms are really understood, but perhaps enough has been said to show how current thinking is going and how formidable both the theoretical and experimental problems are.

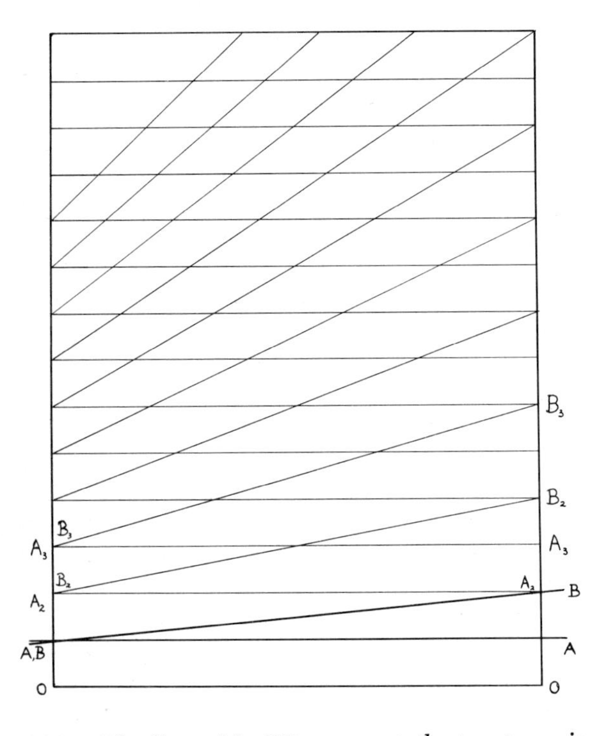

6.16 a *The lines AA, BB represent the two tones in the experiment illustrated in fig. 6.8. The lines A_2A_2, B_2B_2, etc., represent the harmonics or aural harmonics of the two basic notes*

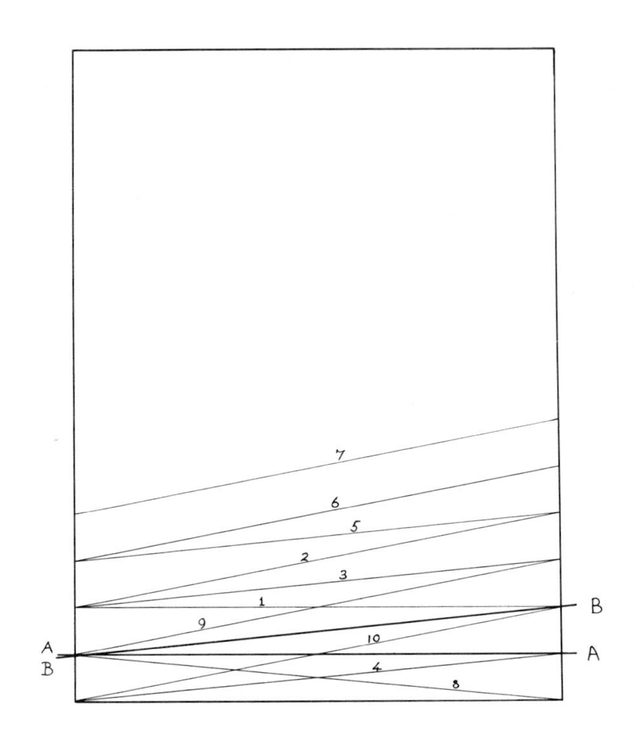

6.16 b *The lines AA and BB represent the two basic notes as in fig. 6.16 a. The other lines represent the following aural harmonics and combination tones. (1) 2A, (2) 2B, (3) A+B, (4) B−A, (5) 2A+B, (6) 2B+A, (7) 2B+2A, (8) 2A−B, (9) 2B−A, (10) 2B−2A*

Some years ago I produced a diagram which has stimulated considerable discussion. I thought I understood it when I drew it, but I now feel that its interpretation is not at all simple! But it is fascinating as a pattern and also seems to suggest a relationship with the problems of consonance, dissonance and hearing, and I therefore reproduce it here. 6.16 a shows how it is built up. It is based on Helmholtz's experiment. The horizontal black line AA represents the steady note and the other thick black line BB represents the second note gliding upwards over an octave. The vertical scale is linear in frequency so that, for example, the lines A_2 A_2, A_3 A_3 represent the successive harmonics of the steady note. The other lines radiating like a fan represent the harmonics of the rising note. 6.16 b shows the next step, which is to add lines representing all kinds of combinations of these

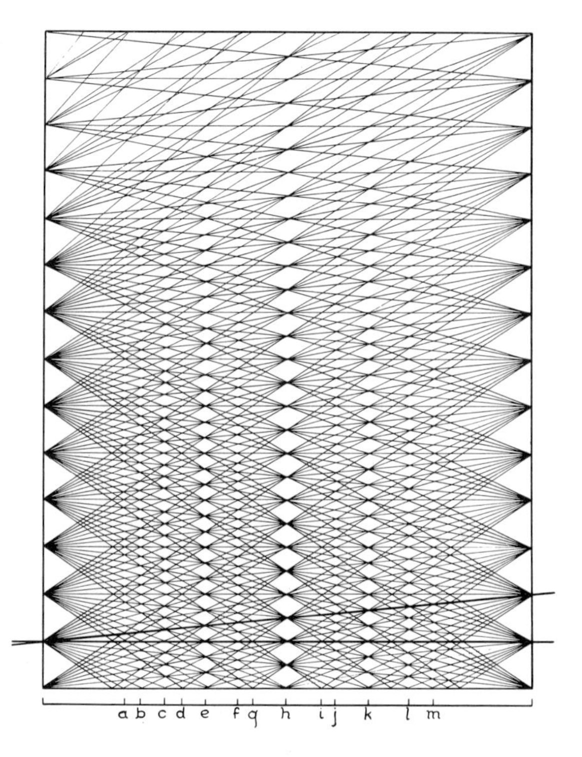

6.16 c *Continuation of the development shown in fig. 6.16 b to include a large number of harmonics and sum and difference tones. The intervals between the basic notes A and B at the lettered points are as follows: (a) Pythagorean minor third (C–E♭). (b) Just minor third (C–E♭). (c) Just major third (C–E). (d) Just diminished fourth (C–F♭). (e) Fourth (C–E). (f) The Pythagorean diminished fifth (C–G♭). (g) Just diminished fifth (C–G♭). (h) Fifth (C–G). (i) Pythagorean minor sixth (C–A♭). (j) Just minor sixth (C–A♭). (k) Just major sixth (C–A). (l) Just augmented sixth (C–A♯). (m) Just minor seventh (C–A♭). It is helpful to view this diagram obliquely with the page tilted backwards.*

notes as detailed in the caption. If this process is continued one eventually arrives at 6.16 c. Without making any claims whatever about the origins of all these additional components, but simply looking at the diagram, one sees that in the vicinity of the consonant intervals – the perfect octave, fifth, etc. – a great simplification occurs and, whatever mechanism is invoked, very few components are present. If one moves just a little bit away from one of these points immediately the number of components increases enormously and one has the discordant, highly complex mixture.

We have spent a long time thinking about harmony and discord. What about musical scales? In many ways of course musical scales are merely methods of classification which make the writing and performance of music simple and give some semblance of a musical language which makes interchange of musical ideas far more feasible than if everyone adopted their own conventions. One could speculate for years on the origin of the various scales used and many books have been written on the subject. Here we shall consider only two systems – that of so-called 'Just Intonation', which tends to be used by string players, and that of 'Equal Temperament', which crops up whenever instruments are used which have the frequencies of their notes determined by the manufacturer rather than by the player – keyboard instruments being the classic examples.

The basic problem is to define a set of frequencies which will blend together in a harmonious way. According to our earlier discussion that means they should give simple frequency ratios when played together. There must not be too many or the system will prove cumbersome, and the steps between the closest ones must be quite clearly distinguishable even by a relatively untrained ear.

The harmonious blending is achieved by using simple ratios in the first place – in other words by using the harmonic series which we would expect from most instruments. The octave (2:1 ratio) and the fifth (3:2 ratio) are obvious starting points and one way to derive the scale of Just Intonation is to take a base note (which we take to be a frequency 1) together with notes of a fifth above and a fifth below and take the first five harmonics of each. Some of these notes will then lie outside the first octave above our base note but they can easily be brought within it by successively dividing or multiplying by 2. Table 6.II shows how this works out.

TABLE 6.II

	1st harmonic	*2nd harmonic*	*3rd harmonic*	*4th harmonic*	*5th harmonic*
Base note	1	2	3	4	5
Divide by 2 successively to bring within the first octave	1	2	$1/2 \times 3$	$1/2 \times 4$	$1/2 \times 1/2 \times 5$
Resulting notes	1	2	3/2	2	5/4
Fifth below base note	2/3	4/3	6/3	8/3	10/3
Divide or multiply by 2 successively to bring within first octave	$2 \times 2/3$	4/3	6/3	$1/2 \times 8/3$	$1/2 \times 10/3$
Resulting notes	4/3	4/3	2	4/3	5/3
Fifth above base note	3/2	6/2	9/2	12/2	15/2
Divide or multiply by 2 successively to bring within first octave	3/2	$1/2 \times 6/2$	$1/2 \times 1/2 \times 9/2$	$1/2 \times 1/2 \times 12/2$	$1/2 \times 1/2 \times 15/2$
Resulting notes	3/2	3/2	9/8	3/2	15/8

Now if we collect up the results and arrange in order we have

$$1 \quad \frac{9}{8} \quad \frac{5}{4} \quad \frac{4}{3} \quad \frac{3}{2} \quad \frac{5}{3} \quad \frac{15}{8} \quad 2$$

and this is the scale which, in essence, forms the basis of conventional music and if we call the base note C we have the succession of white notes on the piano CDEFGABC, the so-called diatonic scale. So far so good – in combination this will certainly lead to mixtures which blend well.

Suppose, however, that we wish to move a little higher in pitch and start on the second note (D). What notes would we need then? All we need to do is to multiply each of the ratios in our diatonic scale by 9/8 and we arrive at

$$\frac{9}{8} \quad \frac{81}{64} \quad \frac{45}{32} \quad \frac{36}{24} \quad \frac{27}{16} \quad \frac{45}{24} \quad \frac{125}{64} \quad \frac{18}{8}$$

$$\text{or} \quad \frac{9}{8} \quad \boxed{\frac{81}{64}} \quad \boxed{\frac{45}{32}} \quad \frac{3}{2} \quad \boxed{\frac{27}{16}} \quad \frac{15}{8} \quad \boxed{\frac{135}{64}} \quad \frac{9}{4}$$

The notes enclosed in a square are new ones which did not occur in our first scale. Suppose we try again on E:

$$\frac{5}{4} \quad \frac{45}{32} \quad \frac{25}{16} \quad \frac{20}{12} \quad \frac{15}{8} \quad \frac{25}{12} \quad \frac{75}{32} \quad \frac{10}{4}$$

$$\text{or} \quad \frac{5}{4} \quad \frac{45}{32} \quad \boxed{\frac{25}{16}} \quad \frac{5}{3} \quad \frac{15}{8} \quad \boxed{\frac{25}{12}} \quad \boxed{\frac{75}{32}} \quad \frac{10}{4}$$

and we have three more notes again. Without going any further one can see that the complications are immense and – though all these variations are easily possible for a string player with a good ear and the necessary playing skill – for a keyboard instrument the number of notes required would be prohibitive. Clearly we need a compromise and that is exactly what Equal Temperament sets out to produce.

If we look at the ratios between successive notes on the just diatonic scale they come out as follows:

$$1 \quad \frac{9}{8} \quad \frac{5}{4} \quad \frac{4}{3} \quad \frac{3}{2} \quad \frac{5}{3} \quad \frac{15}{8} \quad 2$$

$$\frac{9}{8} \quad \frac{10}{9} \quad \frac{16}{15} \quad \frac{9}{8} \quad \frac{10}{9} \quad \frac{9}{8} \quad \frac{16}{15}$$

In other words there are three types of interval, two of which are nearly equal and one of which is almost half as big. They are known as the major tone (9/8), minor tone (10/9) and semitone (16/15). If we make the approximation that major tone equals minor tone equals twice a semitone the octave is then divided into twelve equal semitones. The ratios then turn out to be approximately

1.000, 1.059, 1.122, 1.189, 1.260, 1.335, 1.414, 1.498, 1.587, 1.682, 1.782, 1.888, 2.000.

The intervals for the diatonic scale become

1.000, 1.122, 1.260, 1.335, 1.487, 1.682, 1.888, 2.000

compared with the just scale expressed decimally

1.000, 1.125, 1.250, 1.333, 1.500, 1.666, 1.875, 2.000.

Clearly it is not quite in tune – but now there are no problems if one starts on a different note. As we mentioned in an earlier chapter, Bach's forty-eight preludes and fugues were written largely to demonstrate that playing in all keys is possible and acceptable.

This has been a thumbnail sketch of the problems of scales and harmony, but perhaps enough has been said to show how much ingenuity has gone into this field in the past and perhaps to stimulate the reader to read more in one of the many texts available. In Appendix II we give some details of scales, frequencies, ratios, etc., that may be of value.

CAN YOU BELIEVE YOUR EARS?

So many hints have been dropped that it will be pretty obvious that the answer to the question is almost certain to be 'no'. In fact there are aural illusions just as there are optical illusions, and if this were not so, we should find life a good deal more complicated than it is. Most readers will have listened to music on a transistor radio with a loudspeaker unit probably not more than 2 or 3 inches in diameter. It is highly improbable that this tiny speaker is capable of reproducing at all adequately any of the lower frequencies that make up musical sounds and yet we quite happily listen to a bassoon solo, a double-bass rhythm accompaniment or a bass singer. This is an illusion in something like the same sense that watching colour television is an illusion. Only three colours are actually present on the TV screen, and our eyes and brains combine the dots and create the illusion that we are really watching scenes containing all the possible colours of the rainbow. Similarly, although a surprisingly limited range of audio frequencies is presented by the radio, our brains create the illusion that all frequencies are present.

In speech we can make do with even less information, and if the frequency band is cut down to only about half an octave the voice may not be recognised as belonging to the person who is actually speaking, but the words can be understood quite well. (5.24 in the last chapter relates to this point.)

Aural illusions are not quite so easy to demonstrate as optical ones since they usually need laboratory apparatus, but one can find quite close parallels. There is a well-known optical fatigue effect in which, if one looks at, say, a red patch for a few seconds and then at a white sheet, one immediately sees a blue-green patch. The parallel involves exposing the subject to noise from which certain frequencies have been eliminated. It sounds rather like ordinary white noise, but if real white noise is suddenly switched on in place of the filtered noise, it assumes a pitch – or indeed is often described as 'assuming a colour'. After a few moments the sensation of pitch dies away and the true sound of white noise reappears.

The blending of pure tones provides numerous illusions. For example, if we start with a pure tone of a certain frequency and then add to it successively further pure tones corresponding to the harmonics of the first one up to, say, five components, it is still possible to distinguish the five separate notes. If all five are switched off and immediately switched on again *together* the illusion of separation vanishes in a most magical way, and when the sound returns it is heard as a single tone of rather reedy quality.

Finally in this very brief selection we should

mention the analogue of the barber's pole that has caused a great deal of interest. The barber's pole has a helical stripe running round a cylinder and when the cylinder rotates one has the illusion that the stripe is moving upwards or downwards depending on the direction of rotation. The illusion is strongest if the eye is fixed on a particular stripe. If on the other hand you look at the whole cylinder it is quite obvious that it is not moving up or down at all. It is not easy to produce the aural illusion with conventional instruments, although approximations exist in musical compositions. Using a computer, however, the illusion can be very successfully demonstrated. The sound produced can best be explained in two stages. First imagine a set of pure tones all an octave apart – they might be represented on a logarithmic frequency scale by the vertical lines in 6.17 a. Now imagine that this whole complex is fed into a system which has a formant characteristic like that of 6.17 b. Each individual note will now start imperceptibly at the bottom, glide up, reaching a maximum round about 1000 Hz, and then fade away (6.17 c). The net result is a most infuriating sound which appears to be continually rising in pitch and yet it never seems to get anywhere. The parallel with the optical case is quite close. If you pick out one note and listen to it you can follow it upwards until it fades away – but the overall effect is always of a complex sound round about 1000 Hz and this supplies the complementary effect of no movement.

The point of discussing these illusions, however briefly, is that they really lead the way to our concluding sections.

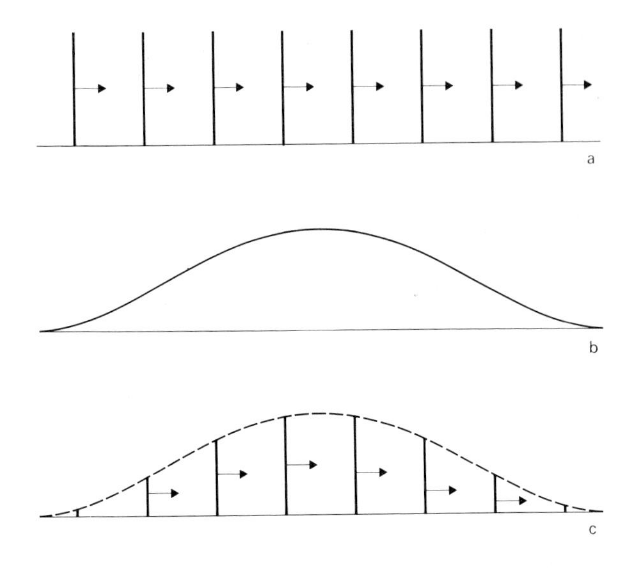

6.17 a–c *Illustration of the principle of the apparently continuously rising tone*

BUT THERE ISN'T ENOUGH INFORMATION

Most of the sound waves which we happily interpret every day of our lives do not appear to carry enough information and yet we make do with it surprisingly well. We saw in the last section that speech can be filtered so that a narrow frequency band only remains and yet it can still be understood. The explanation of the apparent miracle of rapid interpretation performed by the ear and brain begins to reveal itself when you try to put various sorts of waves under very close scrutiny or when you examine very short portions of waves.

An interesting demonstration that illustrates this point can be done by anyone endowed with patience and a tape-recorder. First record a phrase sung by a soprano which includes a fairly high note with a reasonably long vowel sound. Cut out from the tape a short section from the middle of this vowel and splice it into the middle of a length of blank tape. When played back the vowel will almost certainly sound like 'ah' whatever it actually is – because that is the vowel with the simplest formant characteristic; with a high-pitched note the brain does not have much in the way of formants to work on! But when the whole phrase is heard the vowel can be recognised easily; it is then set in a context rather than being examined in isolation. My own demonstration example is from *Dido and Aeneas* in which the soprano sings 'No more' with the 'more' on a very high note. When the 'or' part only is played back it is quite clearly 'ah'!

A remarkable example of the importance of context is demonstrated by the method of sending messages on 'talking drums' which has been used in Africa particularly. The whole process is described in a fascinating book by John Carrington called *The Talking Drums of Africa.* He gives many examples, but one will suffice. A particular drum has two tones, a high and a low one, and words can be imitated by combinations of high and low beats to represent the intonation.

But it is equally clear that in any language there might be a large number of words which tonally could be represented by the same sequence.

Father, mother, brother, sister, to use English examples, would all be the same; elephant, telephone, gramophone would differ from the first group but would have the same tonal pattern as each other. How then can precise messages be sent with only two tones? The answer is that multiple interpretations are gradually eliminated by placing the words in a context and always using standard phrases. For instance, to distinguish between the first four English examples, phrases such as 'Father who is head man of the family', 'Mother the woman who bears children' might be used. Actual examples given by John Carrington in his book and taken from the Kele language of Central Africa are the words for the moon and for a fowl, which have identical tonal patterns. But the phrases used are 'the moon looks down at the earth' and 'the fowl, the little one which says kiokio'.

If you think this sounds complicated, remember that it is really very little different from the use of binary arithmetic in computing where again all the complex messages, instructions, data, etc., are all represented by quantities which have only two possible values symbolised as 0 or 1.

Where is this discussion leading? It forces us to take note of the importance of all the complex interactions and interconnections that occur in our aural-perception apparatus, and this, of course, is precisely why, at the beginning of this chapter, I stressed the weakness and over-simplification of the idea that the ear is just like a telephone converting sound waves into electrical signals. Our ears and brains are ready to make use of all kinds of fragments of information from all sorts of sources – not only in the wave being interpreted, but in comparisons with earlier parts of the wave and with information which is already stored as memory or experience.

FITTING TOGETHER THE CLUES

The whole process is rather like a piece of detective work – but of course the speed at which it happens is the startling feature. There is a machine called a tempophon that enables one to speed up speech without altering its pitch. In effect it is a tape-recorder with several play-back heads on a revolving drum; the drum runs in the same direction as the tape but at slower speed, so the tape moves over the heads at its normal speed while the tape as a whole races through the machine much faster. What really happens then is that the speech is chopped up into tiny bits and we are given every second or every third bit; the brain quite happily smooths over the missing bits and it just appears that the speech is very rapid. At about three to four times normal rate it becomes virtually impossible to understand what is being said unless one is given a clue. For example, if the record is of a well-known nursery rhyme but the listener does not know this, he claims that he cannot understand. If he is then told what it is and listens again he cannot understand how he could possibly have misunderstood the first time! This is typical of brain operation – the slightest clue may set in motion the mechanism that makes everything clear.

To conclude, let us return to the problem with which we began the whole story. The air at a particular point can only have one value of pressure at one point at one time. Look again, for example, at the traces in Appendix IV and see how combinations of several instruments still produce only one wave form. How then can the sound of a whole opera with a choir of a hundred voices, and an orchestra of eighty players, be conveyed to the brain of the listener as one variation of pressure with time? And how can he also hear the clink of the coffee cups being prepared for the interval, the snores of an uninspired music critic and the whispered admonitions of his neighbour, all at the same time, with the ability to concentrate more or less at will on any component of this complex combination? The answer must lie in the vast number of clues that are being given all the time and which can be picked up, identified, compared with previous experiences and interconnected in various ways by the brain all in a split second. The brain can also be persuaded to supply out of its memory and experience various missing items of information. We have said very little about stereophonic sound and the even more recent quadraphonic or multiphonic systems, and of course it is true that with these one is given more information to go on. But no matter how complex our system for actually creating the waves – and a full chorus and orchestra is about as complex as one could wish – we have two ears and hence the basic acoustic information can never be more than two sequences of pressure variation with time.

We have spent most of this book talking about the way in which the original waves in the air are created, about all the little modifications and imperfections that are so valuable in giving clues, about the hazards they undergo on their journey across the intervening space to the ears of the listener, and a little of what happens when they arrive. It must by now be clear, however, that the arrival of the waves at the ear is not by any means the end of the journey but is in fact the beginning of a far more complex and fascinating one through the labyrinth of interconnections in the ear and brain, a partnership about which we really understand very little. Physics plays its part in all the various stages of both journeys but, on its own, it cannot even begin to explain the pleasure, the enjoyment, the challenge, the creative satisfaction and the aesthetic experience that are all involved when we listen to the sounds of music.

The Appendices

APPENDIX I

The chart is intended to show the frequency range of an average human ear and also of various musical instruments. It can only be approximate. At either end of the diagram notes an octave apart are the only ones given. The central octave starting on middle C is, however, given with all the equal-tempered semitones (i.e. both black and white notes on a piano). At the top is the musical notation and at the bottom the frequency, assuming A = 440 Hz. The lowest line of all gives the wavelengths of the notes, assuming a velocity of sound of 331 metres per second – which is something like the average value in the open air.

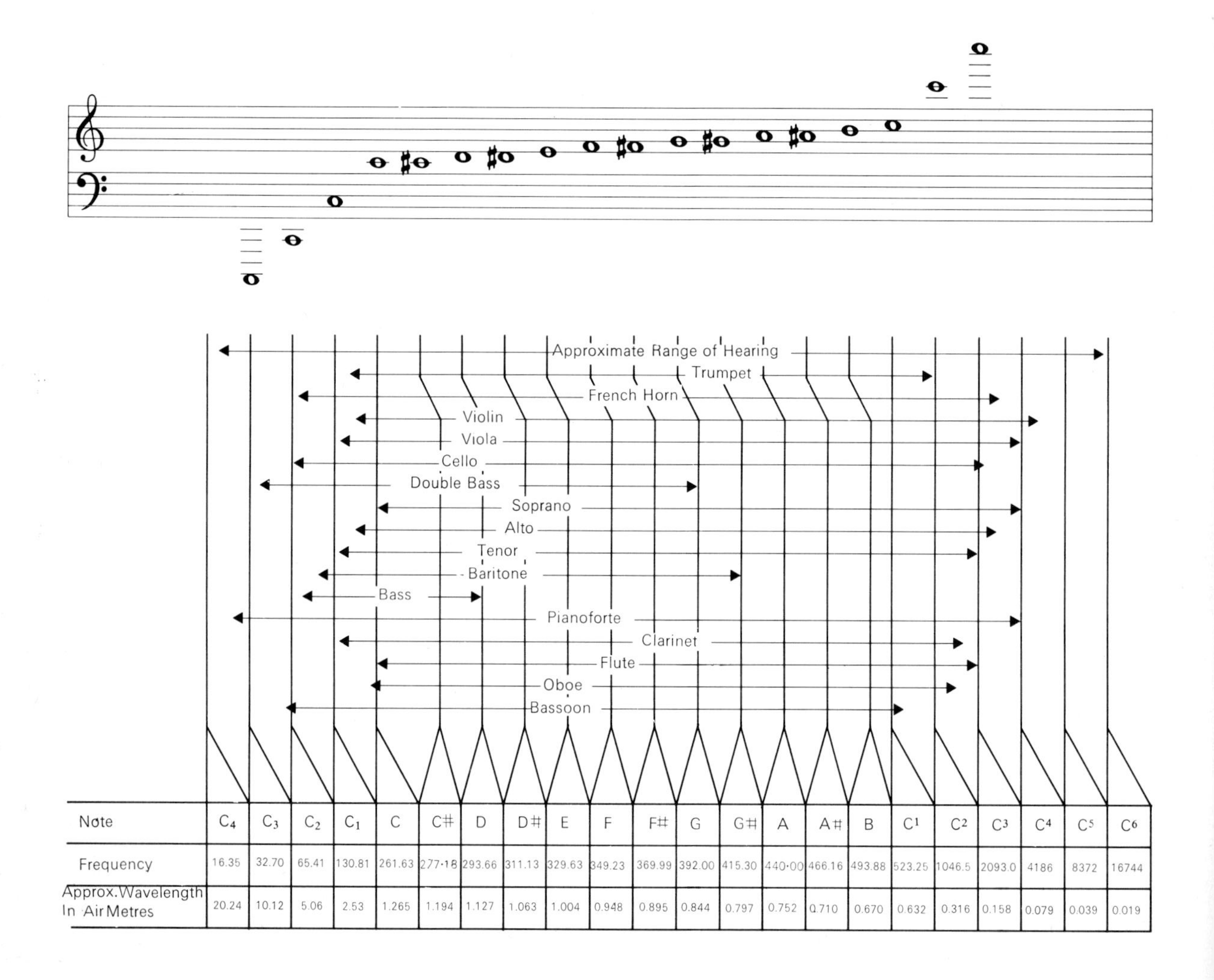

Note	C_4	C_3	C_2	C_1	C	C♯	D	D♯	E	F	F♯
Frequency	16.35	32.70	65.41	130.81	261.63	277·18	293.66	311.13	329.63	349.23	369.99
Approx. Wavelength In Air Metres	20.24	10.12	5.06	2.53	1.265	1.194	1.127	1.063	1.004	0.948	0.895

Note	G	G♯	A	A♯	B	C^1	C^2	C^3	C^4	C^5	C^6
Frequency	392.00	415.30	440·00	466.16	493.88	523.25	1046.5	2093.0	4186	8372	16744
Approx. Wavelength In Air Metres	0.844	0.797	0.752	0.710	0.670	0.632	0.316	0.158	0.079	0.039	0.019

APPENDIX II

A NOTE ON THE SCALE OF JUST INTONATION AND ON THE USE OF CENTS FOR FREQUENCY MEASUREMENT

As discussed in chapter six, the scales of Just Intonation are developed from the notes of the harmonic series and are therefore likely to produce more acceptable harmonies. For instruments – such as the strings – for which infinitesimal variations in pitch are possible, this presents no problem; for instruments such as those with keyboards, however, there are immense problems because the number of notes required to play in tune in all keys is very great. Again, as was discussed in chapter six, the scale of Equal Temperament is the compromise solution usually adopted and in the table comparisons are given.

A further complication exists in that the various forms of major and minor scale would require different notes. All, however, involve various sequences of only three intervals – the major tone, the minor tone and the semitone. The major diatonic scale, for example, has the sequence major tone, minor tone, semitone, major tone, minor tone, major tone, semitone. But the sequence would be different for a minor scale.

In order to provide information in a useful form we give below the frequencies of the major diatonic scale of Just Intonation starting on C but related to the note A of frequency 440 Hz.

In the larger table are given most of the various combinations of ratios that can occur for scales based on C, and the corresponding intervals and frequencies for Just Intonation are included. For comparison purposes the corresponding notes and intervals for the equal-tempered scale also based on A = 440 Hz are given.

The intervals are given in cents. The cent is a measure of frequency on a logarithmic scale obtained by dividing an octave into twelve equal semitones – as in the equal-tempered system – and defining each semitone as 100 cents. Thus the notes on the equal-tempered scale, as for example on the piano keyboard, are simply 100, 200, 300, 400, 500, 600, 700, 800, 900, 1000, 1100 and 1200 cents above the first note. Mathematically speaking the interval between two notes A and B measured in cents is simply

$$1200 \frac{\log_{10} A/B}{\log_{10} 2}.$$

One big advantage is that intervals can simply be added together. For example, the individual intervals, expressed in cents, making up a complete octave always add up to 1200 (e.g. line 6 of the first table); the interval between the notes C and F would be 204 + 182 + 112 = 498 cents.

Note	*C*	*D*	*E*	*F*	*G*	*A*	*B*	*C*
Frequency	264	297	330	352	396	440	495	528
								1
Ratio to first note	1	9/8	5/4	4/3	3/2	5/3	15/8	2/1
Interval to first note in cents	0	204	386	498	702	884	1088	1200
Interval between successive notes		9/8	10/9	16/15	9/8	10/9	9/8	16/15
Interval in cents		204	182	112	204	182	204	112

		JUST INTERVALS			EQUAL-TEMPERED INTERVALS		
Components	*Name of interval from C*	*Frequency ratio to C*	*Interval from C (to nearest cent)*	*Frequency*	*Frequency of nearest equal-tempered note*	*Cents*	*Note*
Unison		1:1	0	264	261.6	0	C
1 semitone	*Semitone*	16:15	112	281.6	277.2	100	C♯
1 minor tone	*Minor tone*	10:9	182	293.3	293.6	200	D
1 major tone	*Major tone*	9:8	204	297.0			
1 minor tone + 1 semitone		32:27	294	312.8	311.1	300	D♯
1 major tone + 1 semitone	*Minor third*	6:5	316	316.8			
1 major tone + 1 minor tone	*Major third*	5:4	386	330.0	329.6	400	E
1 major tone + 1 minor tone + semitone	*Perfect fourth*	4:3	498	352.0	349.2	500	F
2 major tones + 1 minor tone	*Augmented fourth*	45:32	590	371.2	370.0	600	F♯
1 major tone + 1 minor tone + 2 semitones	*Diminished fifth*	64:45	610	375.4			
2 major tones + 1 minor tone + 1 semitone	*Perfect fifth*	3:2	702	396.0	392.0	700	G
2 major tones + 1 minor tone + 2 semitones	*Minor sixth*	8:5	814	422.4	415.3	800	G♯
2 major tones + 2 minor tones 1 semitone	*Major sixth*	5:3	884	440.0	440.0	900	A
Occurs only as the seventh harmonic of the base note	*Harmonic minor seventh*	7:4	969	462.0			
2 major tones + 2 minor tones + 2 semitones	*Grave minor seventh*	16:9	996	469.3	466.2	1000	A♯
3 major tones + 1 minor tone 2 semitones	*Minor seventh*	9:5	1018	475.2			
3 major tones + 2 minor tones + 1 semitone	*Major seventh*	15:8	1088	495.0	493.88	1100	B
3 major tones + 2 minor tones + 2 semitones	*Octave*	2:1	1200	528.0	523.25	1200	C′

APPENDIX III

This table compares the amplitude of the changes in pressure occurring in a sound wave measured either in Newtons per square metre or as fractions of atmospheric pressure with the corresponding rate of flow of the energy falling on a surface in the sound wave measured in millionths of a watt per square metre.

The scale in the middle is the corresponding sound-pressure level measured in decibels. If the sound pressure level in this column were that of a 1000 Hertz note which seemed to the ear to be as loud as the sound being judged, then the middle column would also be the equivalent loudness in phons.

Amplitude of pressure change in the wave		*Sound pressure level*	*Intensity (energy flow in unit time through unit area)*
Newtons per square metre	*Millionths of an Atmosphere*	*Decibels* (dB)	*Millionths of a watt per square metre*
20	200	120	1,000,000
2	20	100	10,000
0.2	2	80	100
0.02	0.2	60	1
0.002	0.02	40	0.01
0.0002	0.002	20	0.0001
0.00002	0.0002	0	0.000001

APPENDIX IV

In order to try to bridge the gap between musicians and scientists, to illustrate some of the points made at various places in the book and to show once again what a remarkable operation is performed by the ear and brain in translating a varying pressure into the marvellous experience of music, here are two comparisons of wave forms with musical scores.

The first is taken from Mozart's Trio for Clarinet, Viola and Piano (K498) – often known as the Kegelstatt Trio. It is part of the trio section of the Second Movement (minuet and trio). The black-on-white trace is a continuous record of forty-nine bars and the music has been written out above it in such a way that the notes fall over the corresponding notes of the oscillograph trace. (Normally, of course, music is written with bars of varying length depending on the number of notes rather than the time involved in order to economise in space.)

Bars 10–15
Bars 16–21
Bars 22–27
Bars 28–33
tr

Bars 34–39

Bars 40–45

Bars 46–50

The white-on-black traces are certain parts expanded in time so that one can see more of the detailed structure of the wave form.

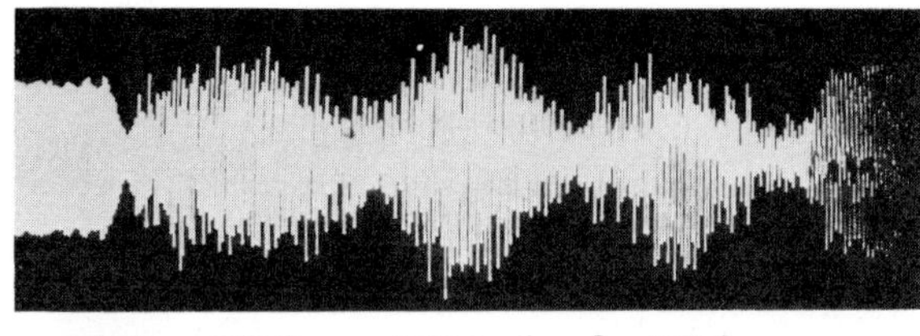

end of note c   note d   note e   2   note f

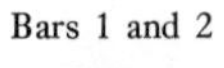

Bars 1 and 2

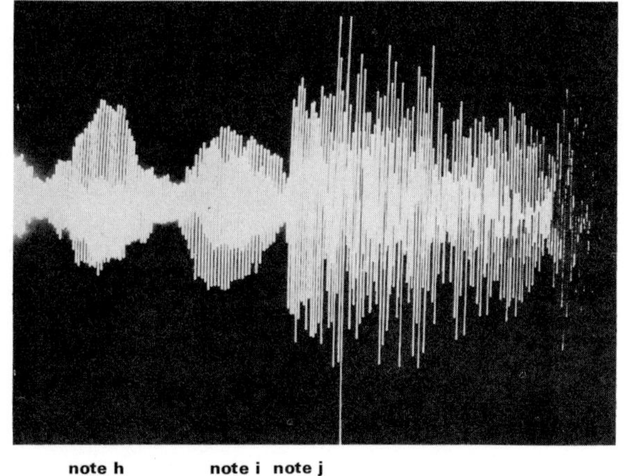

note h viola   note i viola   note j   piano trill

Bar 2

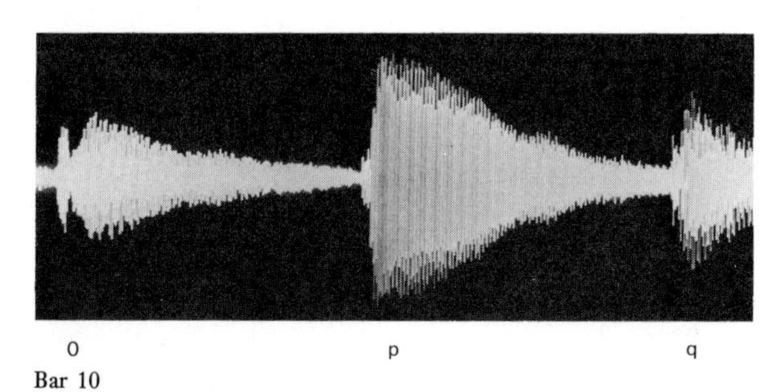

O   p   q

Bar 10

p

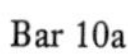

Bar 10a   Note p greatly expanded

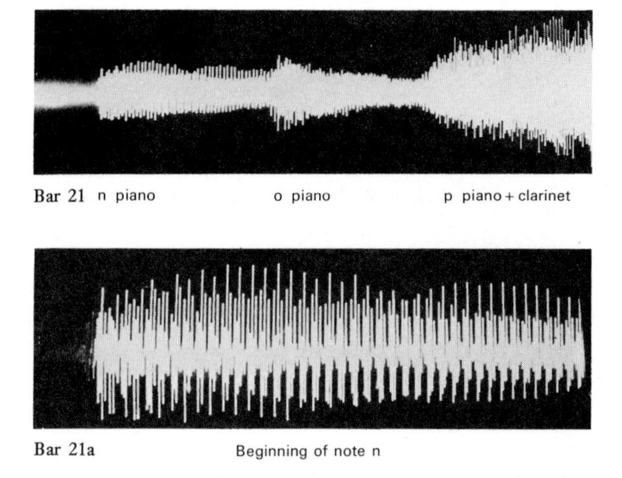

Bar 21   n piano   o piano   p piano + clarinet

Bar 21a   Beginning of note n

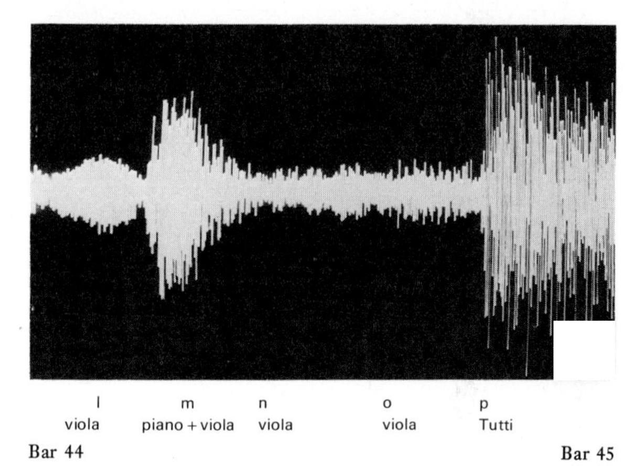

l viola   m piano + viola   n viola   o viola   p Tutti

Bar 44   Bar 45

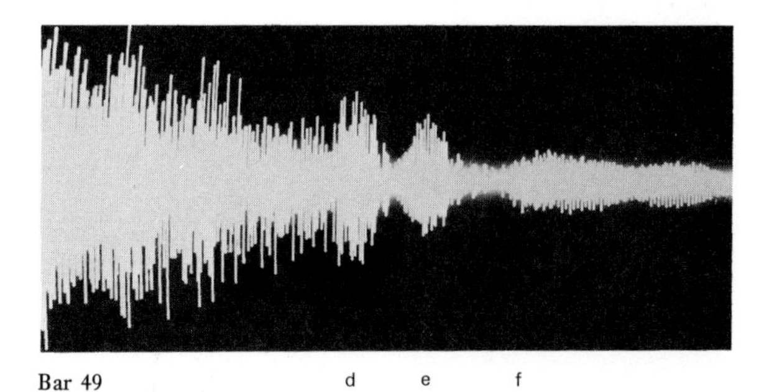

Bar 49   d   e   f

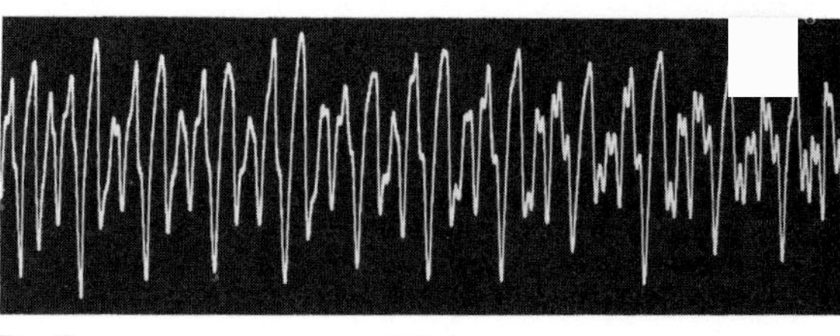

Bar 49a   note f

The second example, below, is nine bars from the Last Movement (allegro) of Mozart's Horn Concerto No. 3 in E♭ major (K47). As before, the music has been rewritten to match the trace, and the wave forms of certain sections have been expanded in time by varying degrees.

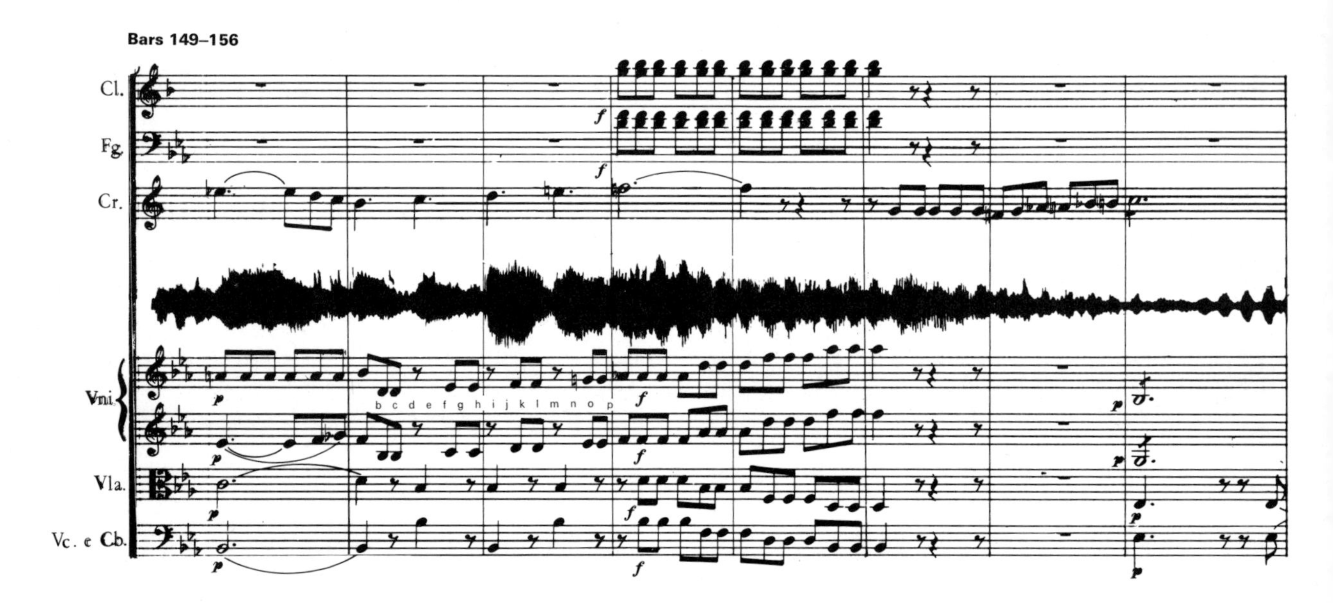

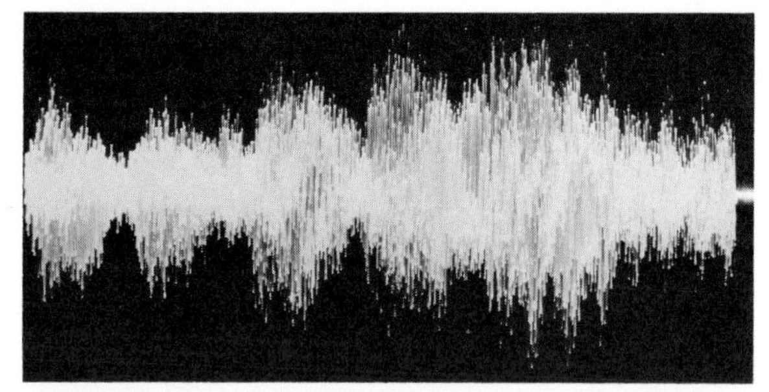

b c d e f g h i j k l m n

Bars 150, 151, 152

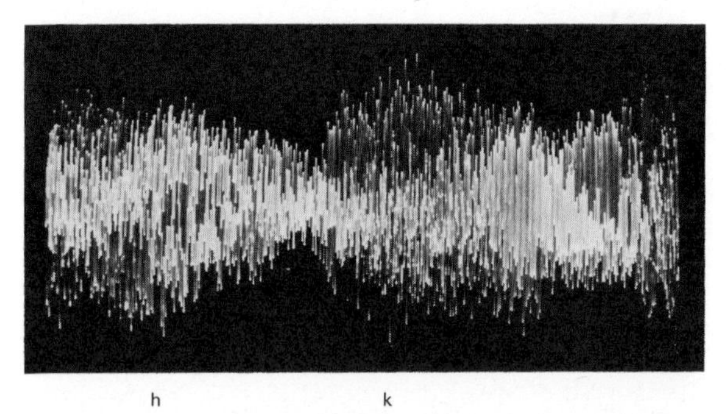

h k

Bar 151

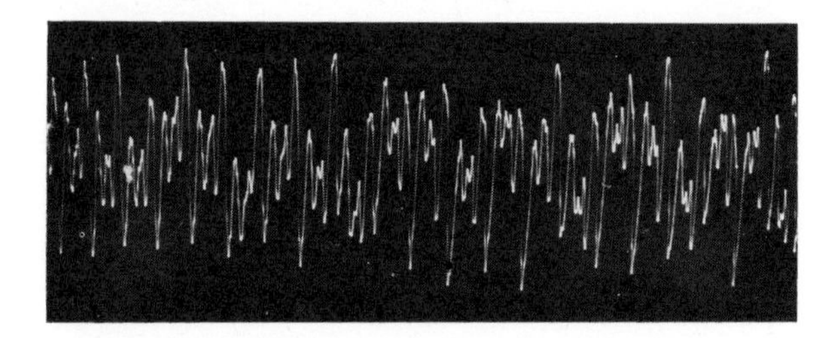

Bar 151 h

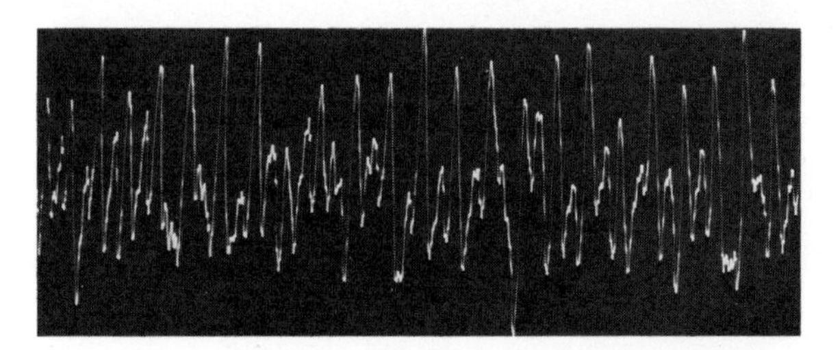

Bar 151 k

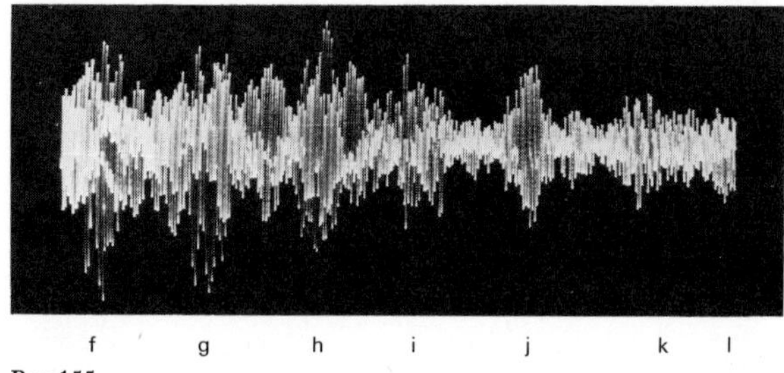

f g h i j k l

Bar 155

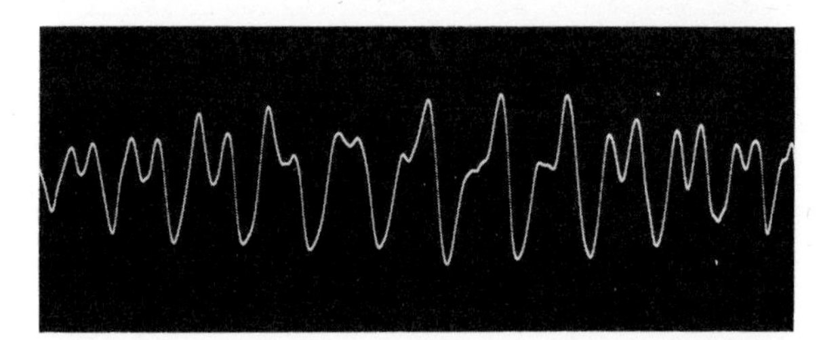

Bar 155 g–h

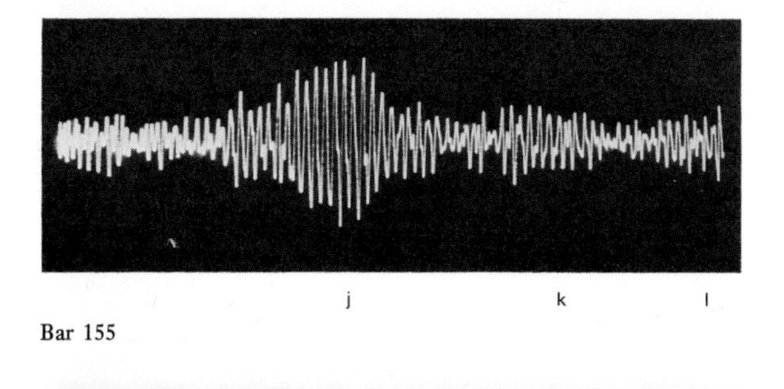

j k l

Bar 155

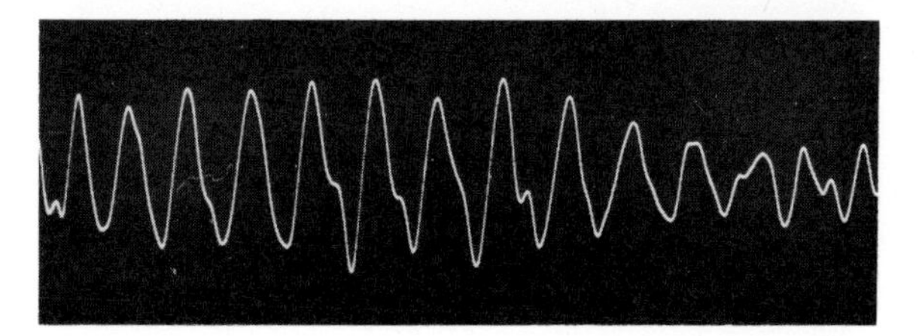

Bar 155 j

l m

Bar 155

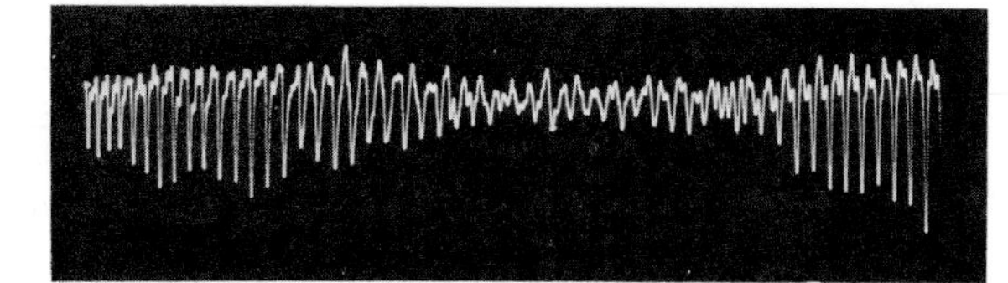

Bar 156 n–o

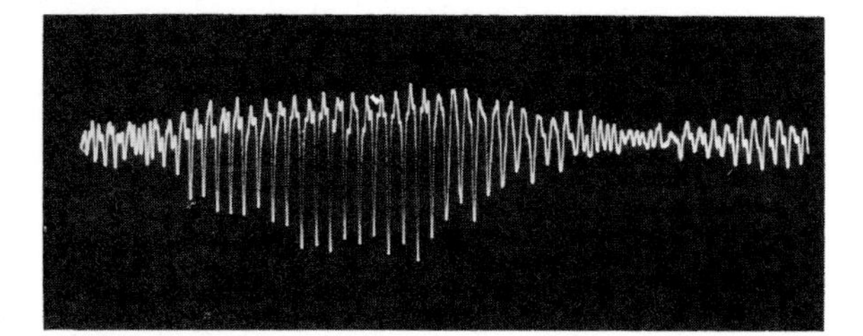

Bar 156 o

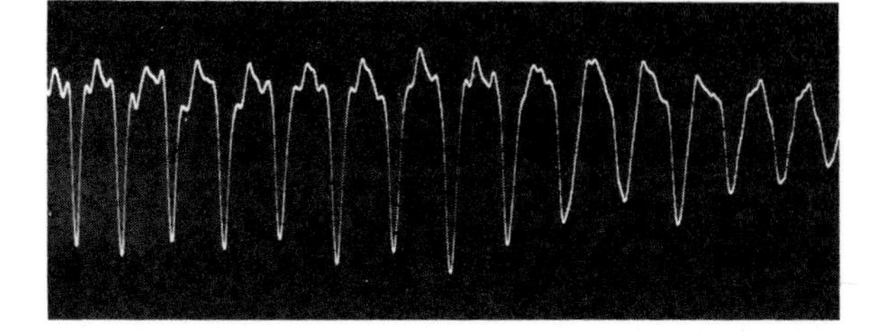

Bar 156 O

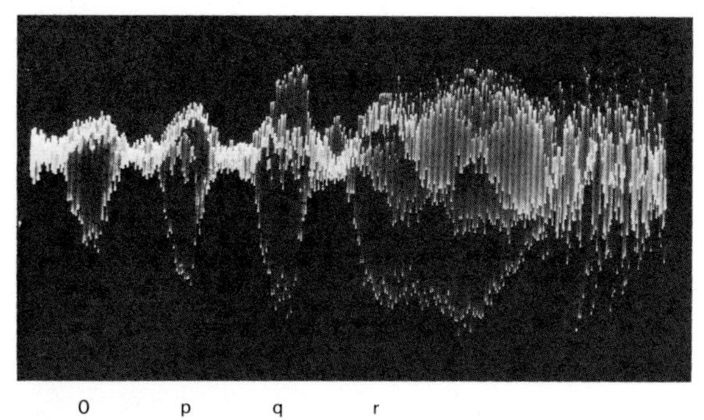

O p q r
Bar 156

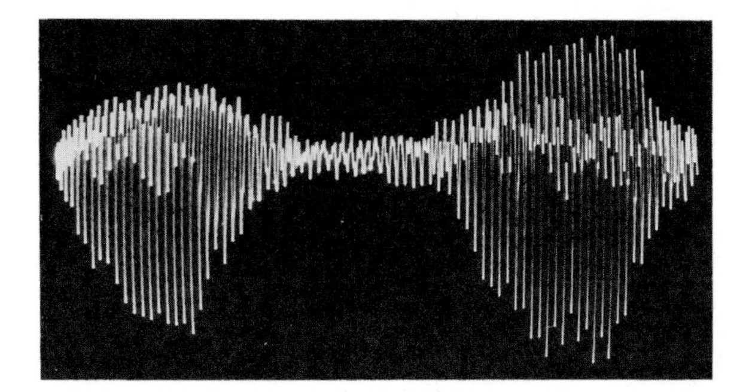

Bar 156 p–q

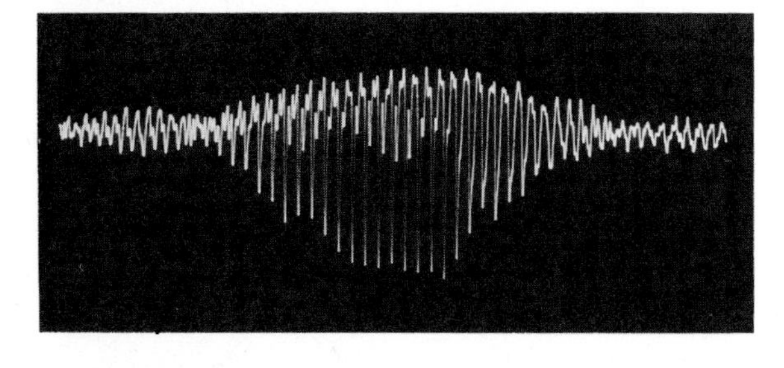

Bar 156 p

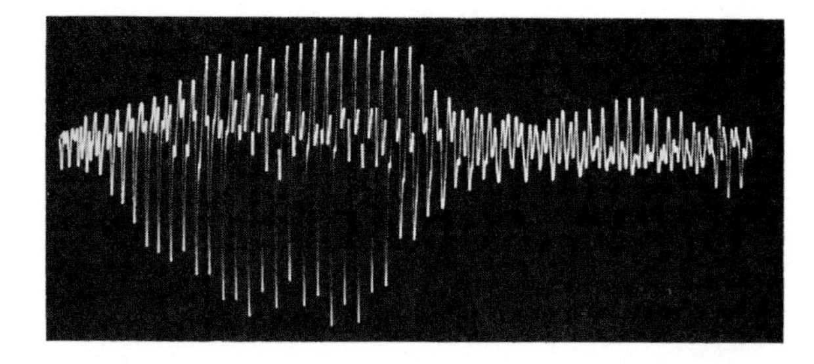

Bar 156 q

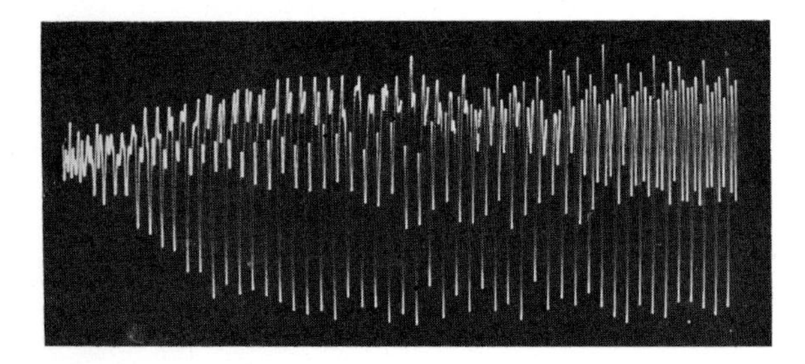

Bar 156 r

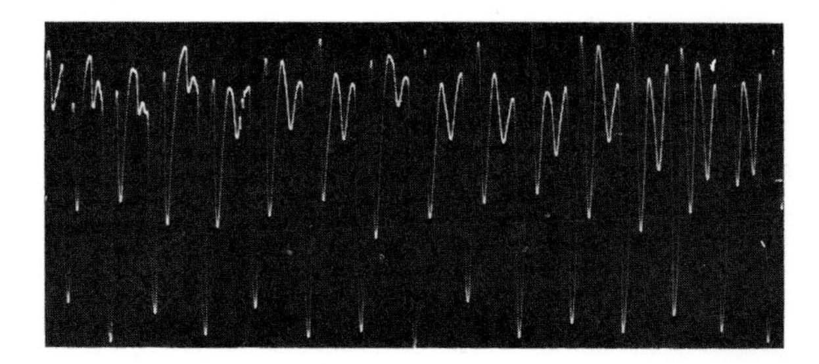

Bar 156 r

SOME SUGGESTIONS FOR FURTHER READING

BACKUS, J. *The acoustical foundations of music* New York, Norton, 1969; J. Murray, 1971.

BENADE, A. H. *Horns, strings and harmony* Doubleday, paperback, 1960. o.p.

BERGEIJK, W. A. van et al. *Waves and the ear* Doubleday, paperback, 1960. o.p.

HELMHOLTZ, H. L. F. *On the sensations of tone* New York: Dover, paperback, reprint of 1875 edn, 1954.

JEANS, SIR JAMES *Science and music* C.U.P., 1937; New York: Dover, paperback, 1968.

JOSEPHS, J. J. *The physics of musical sound* Van Nostrand-Reinhold, paperback, 1967.

LOWERY, H. *The background to music* Hutchison, 1952. o.p.

MILLER, D. C. *Anecdotal history of the science of sound* New York: Macmillan, 1935. o.p.

ROEDERER, J. G. *Introduction to progress and psychophysics of music* Heidelberg Science Library, vol. 16, New York: Springer-Verlag, paperback, 1973.

TAYLOR, C. A. *The physics of musical sounds* English Univ. P., 1966.

TYNDALL, J. *Sound* Longmans, 1867; Dover, 1973

TYNDALL, J. *Heat a mode of motion* Longmans, 1863. o.p.

WINCKEL, F. *Music, sound and sensation* Dover, paperback, 1967.

WOOD, A. *Physics of music* Methuen, 1944, 6th edn, 1962. o.p.

Acknowledgment is due to the following for permission to reproduce illustrations:

Figures: 1.1 a–c, from *Acoustics* by Gerard Hoffnung (Dennis Dobson); 1.1 d, Steinberg; 1.3, by permission of Ernst Eulenburg Ltd; 1.4, by permission of Universal Edition (London) Ltd; 1.5, by permission of Universal Edition (London) Ltd; 1.12, by permission of Ernst Eulenburg Ltd; 1.19, Horniman Museum, London; 1.35, J. E. Bulloz; page 28, J. E. Bulloz; page 52, Staatsbibliothek, Berlin; 3.22, Royal Institution; 4.11 a and b, Horniman Museum, London; 4.11 c, W. E. Hills & Son; 4.18 d, Karl A. Stetson and Catgut Acoustical Society, Inc.; 4.19 a and b, Catgut Acoustical Society, Inc.; 4.21–23, from *The Physics of the Piano* by E. Donnell Blackham. Copyright © December 1965 by Scientific American, Inc. All rights reserved; 4.29, Radio Times Hulton Picture Library; 4.30, *Son* Magazine; p. 122, J. E. Bulloz; p. 124, J. E. Bulloz; 5.11, Royal Institution; 5.23, produced with the permission of the Royal Albert Hall; p. 142, J. E. Bulloz; p. 142, Staatsbibliothek, Berlin; 6.3, from Fundamentals and Acoustics by Kinsler and Frey (John Wiley, 1962); 6.4 and 6.5, from *Fundamentals and Acoustics* by Kinsler and Frey (John Wiley, 1962); 6.6, from *Acoustics* by Gerard Hoffnung (Dennis Dobson), p. 163, Staatsbibliothek, Berlin; 6.8 and 6.12, from *The Physics of Musical Sounds* by C. A. Taylor (English Universities Press).

ACKNOWLEDGEMENTS

A book of this kind evolves over a period of many years and it is virtually impossible to identify all the people who have influenced it in one way or another. I shall try to include as many as possible, but I should also wish to express my gratitude to those whose help and inspiration are not explicitly recorded.

The immediate source of most of the material is the one hundred and forty-second course of Christmas Lectures given at the Royal Institution between 28 December 1971 and 7 January 1972. My first thanks therefore go to Sir George Porter, the Director of the Royal Institution, for inviting me to present the lectures and for giving me the tremendous opportunity of working in the incomparable atmosphere and with the remarkable resources that exist there. The series was televised on BBC2 and various facilities of the BBC were put at my disposal and played a significant part in the success of the enterprise. Mr Alan Sleath, who produced the programmes, shared in the preparations over a period of about 18 months and to him, to the staff of the Royal Institution – especially Mr Bill Coates – and to the members of the BBC production team I owe a great debt of gratitude.

My own staff in the Physics Department at Cardiff contributed in many ways: Mr John Morris and his colleagues in the workshop constructed beautiful apparatus from sketches on the backs of envelopes; Mr Bob Watkins prepared almost all the original photographs and developed not a few new techniques along the way; Dr Philip Williams and Mr Keith Winter, my academic colleagues in the Acoustics and Electronic Music groups respectively, gave a great deal of help and advice and, in particular, Dr Williams read the manuscript; my secretary, Mrs Valerie Chown, typed it.

The Leverhulme Trust gave a substantial grant which enabled us to initiate the cooperative work in research and teaching between the Physics and Music Departments which has since been extended and developed from College resources and became one of the main sources of ideas. The Principal of the College encouraged me to take six months' sabbatical leave during which, among other tasks, a large part of the manuscript was prepared.

During the early stages I visited many studios and laboratories and was delighted with the readiness of so many people to help both with advice and with demonstration material. Among these were Professor Schouten of Eindhoven, Professor Benade of Case Western Reserve University, Dr Max Matthews of Bell Telephone Laboratories, Dr Sundberg of the Royal Institute of Technology, Stockholm, Professor Plomp of Soosterberg, Dr Blackburn and his colleagues at the Plessey Company, Professor Donald Mackay and his colleagues at Keele, Mr Macfadyen of Birmingham, the BBC research centre at Kingswood Warren, the BBC Radiophonic Workshop, Mrs Jean Jenkins of the Horniman Museum, M. F. Baschet of Les Structures Sonores, Mr Peter Zinoviev, Mr Kuehn of B & K, Dr Karl Stetson and his colleagues at the NPL, Mr David Gibbs of Bristol, Dr Ian Firth of St Andrews and Mr Kennedy of Boosey and Hawkes.

Dr John Robertson of Leeds suggested the idea on which Appendix 4 is based; Mr Alec McCurdy showed me his work on 'cellos and allowed us to photograph him at work; Mrs Carleen Hutchins provided a good deal of encouragement and advice and read parts of the manuscript.

Mr Peter Campbell and his colleagues of BBC Publications have in my view done a splendid job of fitting together a large collection of rough sketches, photographs and textual material and creating out of them an attractive layout, and in many ways made the task of preparing the book an interesting and rewarding exercise. And finally my wife and family have given their usual ungrudging and patient support and encouragement. My grateful thanks go to all these and many more.

Index

INDEX

DEC 8 '77